Ordinary Differential Equations
for Engineering and Science Students

Mathematical Topics for Engineering and Science Students
General Editor: Dr S. Barnett, University of Bradford

Other titles in this series

Mathematical Formulae (2nd edition)
S. Barnett and T. M. Cronin

Basic Statistical Techniques
Keith D. C. Stoodley

Elementary Calculus
T. M. Cronin

Ordinary Differential Equations for Engineering and Science Students

L. B. JONES

Bradford University Press
in association with
Crosby Lockwood Staples   London

Granada Publishing Limited
First published in Great Britain 1976 by Bradford University Press
in association with Crosby Lockwood Staples Frogmore St Albans
Hertfordshire AL2 2NF and 3 Upper James Street London W1R 4BP

Copyright © 1976 by L. B. Jones

All rights reserved. No part of this publication may be
reproduced, stored in a retrieval system, or transmitted,
in any form or by any means, electronic, mechanical,
photocopying, recording or otherwise, without the prior
permission of the publishers

ISBN 0 258 96973 3 (hardback)
     0 258 97047 2 (paperback)

Printed in Great Britain by William Clowes & Sons Limited
London, Colchester and Beccles

Contents

Preface   ix

Chapter 1   **Introductory ideas**
1.1   Introduction   1
1.2   Formulation of differential equations   4
1.3   Classification of differential equations   16
1.4   Solution of differential equations   17
1.5   Existence and uniqueness   30

Chapter 2   **Analytic solutions of first-order differential equations**
2.1   Introductory discussion   33
2.2   Separable equations   33
2.3   Homogeneous equations   36
2.4   Equations of the form $\dfrac{dy}{dx} = \dfrac{ax + by + c}{\alpha x + \beta y + \gamma}$ 37
2.5   Exact differential equations   40
2.6   Linear equations   43
2.7   Bernoulli's equation   46
2.8   Miscellaneous examples   48

Chapter 3   **Numerical solutions of first-order differential equations**
3.1   Introductory discussion   54

vi Contents

	3.2	Integration formulae	54
	3.3	One-step methods	62
	3.4	Predictor–corrector methods	68
	3.5	Ill-conditioning, stability	73

Chapter 4 Analytic solution of second- and higher-order differential equations
	4.1	Introduction	79
	4.2	Homogeneous second-order linear equations with constant coefficients	84
	4.3	General second-order linear equations with constant coefficients	88
	4.4	Higher-order linear equations with constant coefficients	99
	4.5	Systems of linear equations with constant coefficients	100
	4.6	Other second- and higher-order equations	109

Chapter 5 Numerical solutions of second- and higher-order differential equations
	5.1	Introductory discussion	113
	5.2	Initial-value problems	114
	5.3	Boundary-value problems	121
	5.4	Stiff equations	126

Chapter 6 Series solution
	6.1	Introduction	130
	6.2	Solution by Maclaurin, Taylor series	130
	6.3	General solution in series about an ordinary point	136
	6.4	General solution in series about a regular singular point	140
	6.5	Special differential equations	158

Chapter 7 Laplace transforms
	7.1	Introduction	165
	7.2	The Laplace transform	167
	7.3	Rules for Laplace transforms	168
	7.4	The Heaviside unit step function	177
	7.5	The delta function	182
	7.6	The inverse Laplace transform	184
	7.7	Solution of linear differential equations	190

Appendix
   Table A   General Laplace transforms   204
   Table B   Particular transform pairs   205

Bibliography   206

Answers to problems   207

Index   219

Preface

The material presented in this book is based on lecture courses delivered by myself to engineering and science students over a period of years. It is hoped that the material presented will provide the reader with a good introductory knowledge of the techniques of analytic and numerical solution of ordinary differential equations. Any student wishing to pursue any part of the subject in further depth can refer to books listed in the short bibliography at the end of this book. Most of the illustrative examples have physical applications and are drawn from various fields. The problems which are given for solution by the student are however given in mathematical form, so that they are of general application rather than being restricted to any one branch of engineering or science. Many problems are given, and answers to all are listed at the end of the book.

In this book when, for example, integrals need to be evaluated reference is made to particular formulae in Barnett and Cronin (1975).

I should like to thank Dr S. Barnett for his many helpful suggestions during the preparation of the manuscript, and for invaluable help in correcting the proofs. Also to Dr J. A. Grant and Mr G. Eccles for their helpful suggestions, to Mrs M. B. Balmforth, Mrs J. Foster and Miss V. M. Morton for typing the manuscript, to Mrs J. Braithwaite and Mr S. Teal for preparing the drawings, and the University of Bradford for permission to make use of examples from University examination papers in preparing the examples and problems of this book.

L. B. Jones
September, 1975

CHAPTER ONE

Introductory Ideas

1.1 Introduction

A differential equation is an equation that involves derivatives. The following are examples of differential equations:

$$\frac{dy}{dx} = x \tag{1.1}$$

$$\frac{d^2y}{dx^2} + y = \sin x \tag{1.2}$$

$$\frac{d^2y}{dt^2} + \left(\frac{dy}{dt}\right)^2 + \sin y = 0 \tag{1.3}$$

$$\frac{\partial^2 y}{\partial x^2} = \frac{1}{c^2} \frac{\partial^2 y}{\partial t^2} \; . \tag{1.4}$$

Equations (1.1)–(1.3) involve only ordinary derivatives, and they are called *ordinary differential equations*. Equation (1.4) involves partial derivatives and so is called a *partial differential equation*.

As the title of the book states, we will be concerned only with ordinary differential equations. Throughout this book, derivatives will be denoted in various ways. If y is a function of x, and assuming that the function is differentiable to whatever order is required, then the first derivative will be denoted by any of the following forms:

$$\frac{dy}{dx}, y', y^{(1)}$$

2   Ordinary Differential Equations

the second derivative by

$$\frac{d^2y}{dx^2}, y'', y^{(2)}$$

and the nth derivative by

$$\frac{d^n y}{dx^n}, y^{(n)}$$

Equation (1.3) can therefore be expressed in the following alternative forms:

$$y'' + (y')^2 + \sin y = 0$$
$$y^{(2)} + [y^{(1)}]^2 + \sin y = 0$$

Given a differential equation involving y, x and the derivatives $d^n y/dx^n$, we wish to determine the resulting dependence of y on x. That is, we wish to find a relationship between y and x, say of the form $y = f(x)$, such that the relationship satisfies the differential equation. Such a relationship is called *a solution* of the differential equation.

For example, by straightforward differentiation it is easily seen that

$$y = \tfrac{1}{2}x^2, \; y = \tfrac{1}{2}x^2 - 1, \; y = \tfrac{1}{2}x^2 + 2$$

are all solutions to the differential equation (1.1), while

$$y = x, \; y = \tfrac{1}{3}x^2, \; y = \tfrac{1}{3}x^3 + 1$$

are not solutions.

Solutions to differential equations must often satisfy certain additional conditions. For example, the path of a projectile may be required subject to its passing through a given point in space with a given velocity at that point; or the deflection of a beam under load may be required subject to its being clamped at two or more places. If all the conditions are prescribed at a given point, as for the projectile, they are called *initial conditions*. If they are prescribed at different points, as for the beam, they are called *boundary conditions*.

Solutions may be obtained in closed form involving known functions, when they are called *analytic solutions,* or they may be obtained as *approximate solutions* which are for practical purposes close enough to the exact analytic solution over a specified range of values; or they may

Introductory Ideas 3

be obtained as *numerical solutions*, usually using a computer. In some cases, however, it may be sufficient to determine only the main characteristics of the solution, rather than the solution itself.

Example 1.1 Consider the differential equation

$$\frac{dx}{dt} = \tfrac{1}{2}(1 - x^2) \quad \text{subject to} \quad x = 0 \text{ when } t = 0 \quad (1.5)$$

x can be taken as the displacement of a particle and t as time.

(a) The analytic solution, which will be obtained in Problem 2.2(a), is

$$x(t) = \frac{e^t - 1}{e^t + 1} \quad (1.6)$$

This can be checked by direct substitution into the differential equation, and can also be seen to satisfy the given condition.

(b) An approximate solution valid for small values of time is

$$x = \tfrac{1}{2}t - \tfrac{1}{24}t^3$$

which is obtained by solution in series as described in Chapter 6. Terms involving t^n where $n \geqslant 5$ have been neglected, and this gives a measure of the accuracy of the approximation.

(c) Using a fourth-order Runge–Kutta method, as described in Section 3.3.2, the solution is

t	0	0·1	0·2	0·3
x	0	0·0500	0·0997	0·1489

where the values of x as given by this numerical method are accurate to about $\pm 10^{-4}$.

(d) The main characteristics of the motion given by (1.5) can be obtained by noting that when $dx/dt > 0$, x increases as t increases, and that when $dx/dt < 0$, x decreases when t increases. Hence for the differential equation (1.5), x increases with time when $-1 < x < 1$ and x decreases with time when $x > 1$ and $x < -1$. When $x = \pm 1$ the velocity of the particle is zero so that the points $x = \pm 1$ are equilibrium points, that is points at which the particle can remain at rest for all values of time. This information is illustrated in Fig. 1.1, the arrows denoting the motion as time increases, starting at various initial points. It will be noted that $x = 1$ is a stable point in that any particle on either side of $x = 1$ approaches $x = 1$, while $x = -1$ is an unstable point.

Fig. 1.1

Returning to the given condition $x = 0$ when $t = 0$, we see from Fig. 1.1 that x will increase with time and eventually approach $x = 1$. This is confirmed by the analytic solution, since from (1.6)

$$\lim_{t \to \infty} x(t) = 1 \quad \text{and} \quad 0 \leqslant x(t) < 1, \quad t \geqslant 0$$

It will be shown in the next section how differential equations arise, and that in most physical situations the differential equation which models a situation is based on a number of approximating assumptions. Accurate solutions of the differential equations are needed when the question of the validity of the model arises, since any discrepancy between the theoretical and experimental results needs to be attributed to the model and not to its solution. When the model is known to be a good representation of the physical situation, then the scientist or engineer may from practical considerations only require the solution to be accurate within prescribed limits. If an analytic solution cannot be found, then any numerical or approximate solutions need to have an accuracy within those prescribed limits. It is therefore necessary in any numerical or approximate solution to be able to quote the order of the error.

1.2 Formulation of differential equations

Differential equations arise in many ways, but we shall mainly be interested in those that result from the mathematical representation of physical situations.

Let us consider in a little detail the mathematical representation of the path of a projectile. The horizontal and vertical axes are denoted by x and y respectively (Fig. 1.2). The projectile P has mass m and is projected with speed V at an angle α to the horizontal.

The only force acting on the projectile is the constant force of gravity, mg, acting in a vertical (downward) direction. Velocity is the rate of change of distance travelled, so that the components of the velocity of the projectile in the x and y directions are dx/dt and

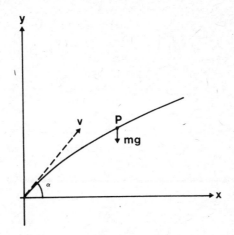

Fig. 1.2

dy/dt respectively. Similarly the components of the acceleration in the x and y directions are d^2x/dt^2 and d^2y/dt^2 respectively.

The differential equations representing the path of the projectile are now obtained by using Newton's law of motion, which states that force equals mass multiplied by acceleration.

Hence in the x direction,

$$m\frac{d^2x}{dt^2} = 0 \qquad (1.7)$$

and in the y direction

$$m\frac{d^2y}{dt^2} = -mg$$

The above pair of differential equations together with the initial conditions

$$\left.\begin{array}{c} x = y = 0 \\ dx/dt = V\cos\alpha \qquad dy/dt = V\sin\alpha \end{array}\right\} t = 0$$

provide a mathematical model of the motion. The solution of the differential equations with the initial conditions results in a relation between y and x which is the equation for the path of the projectile. This equation is obtained later in Problem 4.9, and is the equation of a parabola, as is well known. This parabolic path is the exact solution of the differential equations, but we now question whether this path coincides with or approximates closely the actual path of the projectile. In other words, is the mathematical model an accurate enough representation of the actual physical situation? We look at some of the assumptions or approximations that have been made to obtain our mathematical model.

(a) We have assumed that the gravitational force is constant. In fact the force is a function of its position above the earth. Taking the earth to be of constant density of radius R, then the gravitational acceleration g at height h above the earth is more closely given by

$$g = \frac{g_0 R^2}{(h + R)^2}$$

where g_0 is the gravitational acceleration at the earth's surface. When $h \ll R$, the value of g can be approximated by g_0, so that when we are concerned with the trajectory of a bullet, golf ball or even flight of an aircraft, the gravitational force can be approximated by a constant value. When we are dealing with the motion of a satellite, the gravitational force has to be taken as obeying an inverse square law, as above. Assuming that the gravitational force due to other bodies such as the sun is small compared with that of the earth, the motion is an orbit with the centre of the earth as focus. Strictly, even when dealing with motion near the earth's surface the path is an orbit with the earth's centre as focus, but this orbit over the small range of distances involved is very closely approximated by the parabolic path obtained on taking a constant gravitational force.

(b) We have assumed that the only force acting on the projectile is that due to gravitation. It is very likely, however, that other forces act on the projectile and that they are not negligible in comparison with the gravitational force. For instance, if we consider motion through air, there is a force resisting motion due to friction between the air and the projectile, and there might be other forces depending upon the shape of the projectile: for example, wings are designed to provide a lifting force (force in the vertical direction). These forces for a particular projectile have to be obtained from aerodynamic theory.

We now consider the case in which the resultant of the forces, other than gravitational, acting on the projectile opposes its motion and is proportional to its velocity (Fig. 1.3). Let v be the speed of the projectile and θ be the angle that the tangent to the path makes with the horizontal.

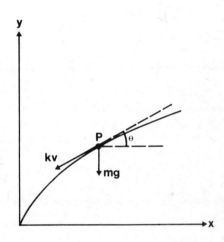

Fig. 1.3

The component of force in the x direction is $-kv\cos\theta$, and in the y direction is $-kv\sin\theta$. But $v\cos\theta$ is the component of the velocity in the direction of the x axis, and hence $v\cos\theta = \mathrm{d}x/\mathrm{d}t$, and similarly $v\sin\theta = \mathrm{d}y/\mathrm{d}t$.

The differential equations giving the motion are now

$$m\frac{\mathrm{d}^2 x}{\mathrm{d}t^2} = -k\frac{\mathrm{d}x}{\mathrm{d}t}$$
$$m\frac{\mathrm{d}^2 y}{\mathrm{d}t^2} = -k\frac{\mathrm{d}y}{\mathrm{d}t} - mg$$

(1.8)

with the same initial conditions as taken previously. Equations (1.8) are the model needed when the forces are as stated. However, the air forces acting on the projectile need not lie in the x,y plane, and if they do not, the path of the projectile is not two dimensional. A set of three differential equations is needed to represent the projectile path. An example of this is well known to all golfers when a certain spin imparted to a golf ball results in a slice or hook shot.

We see from the above discussion that our original set of differential equations is an adequate approximation to the physical situation, provided that the projectile stays close to the earth's surface and that air or other forces are of very small magnitude in comparison to the gravitational force. If any of these conditions are not satisfied, then a different set of differential equations is needed.

It must therefore be kept in mind that representation of the physical situation by a set of differential equations is only an approximation, albeit a valid approximation, to the real case. This arises because the basic physical 'law' may be valid only over a certain domain of the variables, and because expressions for quantities, such as forces, may also be valid only over a certain domain. However, within the assumptions made, the model is an exact representation of an idealised physical situation.

A few examples follow. The 'laws' used are quoted and the mathematical model obtained. One assumption in each example is commented upon, and the student should think about any other assumptions that have been made in setting up the model, and also about the validity of the 'laws' quoted, consulting, when necessary, books with the required physical background.

1.2.1 Cooling of a hot body

A body initially at temperature T_0 is surrounded by a medium at constant temperature T_m ($T_0 > T_m$). The cooling of the body is taken to be governed by Newton's law of cooling, which states that the rate of decrease of the body temperature is proportional to the difference between the body temperature and that of the surrounding medium.

Let T be the temperature of the body at time t.

Rate of decrease of body temperature $= - \mathrm{d}T/\mathrm{d}t$.

The minus sign occurs since $\mathrm{d}T/\mathrm{d}t$ is the rate of increase of temperature with respect to time.

The excess temperature $= T - T_m$.
Hence

$$-\frac{dT}{dt} = k(T - T_m) \tag{1.9}$$

where $k\ (>0)$ is the constant of proportionality.

This is the differential equation governing the cooling of the body and is subject to the *initial condition*

$$T = T_0 \quad \text{when } t = 0$$

(At any time t, all points of the body are assumed to have the same temperature T.)

1.2.2 A two-mass-spring system

The system is hanging vertically as shown in Fig. 1.4. The two similar weightless springs have natural length L and modulus of elasticity λ.

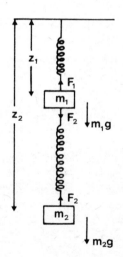

Fig. 1.4

10 Ordinary Differential Equations

The restoring force of a spring is assumed to be given by Hooke's law, which states that this force is proportional to the extension of the spring.

Let z_1 and z_2 be the distances of the masses m_1 and m_2 from the point of suspension of the system.

Let F_1 and F_2 be the restoring forces of the upper and lower springs respectively.

Then using Newton's law of motion, the motion of mass m_1 is governed by the differential equation

$$m_1 \frac{d^2 z_1}{dt^2} = m_1 g - F_1 + F_2$$

and that of m_2 is

$$m_2 \frac{d^2 z_2}{dt^2} = m_2 g - F_2$$

The forces F_1 and F_2 are given by Hooke's law as

$$F_1 = \frac{\lambda(z_1 - L)}{L} \qquad F_2 = \frac{\lambda(z_2 - z_1 - L)}{L}$$

Hence the pair of differential equations governing the motion of the system is

$$m_1 \frac{d^2 z_1}{dt^2} + \frac{2\lambda z_1}{L} - \frac{\lambda z_2}{L} = m_1 g$$

$$m_2 \frac{d^2 z_2}{dt^2} - \frac{\lambda z_1}{L} + \frac{\lambda z_2}{L} = m_2 g + \lambda$$

(1.10)

This will be subject to the initial conditions appropriate to the particular motion being investigated.

(Hooke's law is valid only for a restricted range of small extensions.)

We now consider the case $m_2 = 0$, so that we are in effect dealing with a one-mass–spring system.

Equation (1.10) reduces to

$$m_1 \frac{d^2 z_1}{dt^2} + \frac{\lambda z_1}{L} = m_1 g + \lambda$$

(1.11)

$$z_2 = z_1 + L$$

The second equation states that the lower spring moves up and down without extension, as is to be expected.

When the system is at rest, the distance of the spring from the upper fixed position is given by

$$z_{equilibrium} = L(m_1 g + \lambda)/\lambda$$

If we take y to be the position of the mass from its equilibrium position, then

$$z_1 = y + L(m_1 g + \lambda)/\lambda$$

and if this is substituted into (1.11), the resulting differential equation in y is

$$\frac{d^2 y}{dt^2} + \frac{\lambda}{m_1 L} y = 0 \qquad (1.12)$$

Further, if we define non-dimensional quantities x and τ given by

$$y = Lx \quad \text{and} \quad t = \sqrt{(m_1 L/\lambda)}\,\tau$$

the differential equation in x and τ is

$$\frac{d^2 x}{d\tau^2} + x = 0 \qquad (1.13)$$

1.2.3 Flow of liquid from a spherical tank

A spherical tank of radius R has an outlet of small area A at the bottom.

Torricelli's law states that the velocity, v, with which the liquid issues from the outlet is $v = 0 \cdot 6\sqrt{(2gh)}$, where h is the depth of the liquid in the tank (Fig. 1.5).

The rate of volume of liquid leaving the vessel is $vA = 0 \cdot 6\sqrt{(2gh)}\,A$.

The volume of liquid in the tank is $\frac{1}{3}\pi h^2 (3R - h)$; hence the rate of *decrease* of volume of liquid in the tank is

$$-\frac{d}{dt}\left[\frac{\pi}{3} h^2 (3R - h)\right] = -\pi \left[2Rh - h^2\right] \frac{dh}{dt}$$

This rate of decrease must equal the rate of volume of liquid flowing from the tank; hence

$$0 \cdot 6\sqrt{(2gh)}\,A = -\pi(2Rh - h^2)\,dh/dt$$

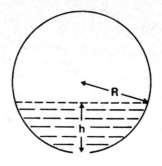

Fig. 1.5

that is,

$$\sqrt{h}\,(h - 2R)\,\frac{dh}{dt} = \frac{0\cdot 6\sqrt{(2g)}A}{\pi} \qquad (1.14)$$

(The radius of the outlet is assumed small compared with the radius of tank.)

1.2.4 Spread of an infectious disease

We consider an isolated community of N people, and make the following assumptions:

   Any person having had the infection is immune from any further infection.
   The infected people recover at a given rate.
   Any uninfected person is liable to be infected by an infected person, and the probability of infection in a given time is a known constant.

Let $U(t)$ be the number of uninfected people at time t,
   $I(t)$ be the number of infected people,
   $R(t)$ be the number of recovered people who are immune.

We note that

$$N = U + I + R$$

Introductory Ideas 13

The differential equations governing the infection are

$$\frac{dR}{dt} = \alpha I \qquad \frac{dU}{dt} = -\beta UI \qquad \frac{dI}{dt} = \beta UI - \alpha I \qquad (1.15)$$

The first equation states that the rate at which the number of recovered people increases is proportional to the number of infected people. This follows since each infected person recovers in a given fixed time. The second equation states that the rate at which the number of uninfected people decreases is proportional to the product of uninfected and infected people. This follows since at a given time each of the I infected people can infect each of the U uninfected people. The third equation follows easily from the fact that $(U + I + R)$ is a constant and hence

$$d(U + I + R)/dt = 0$$

This set of differential equations has to be solved subject to U, I and R having given values at time $t = 0$.

(Outside factors such as immunisation are not taken into account.)

We now turn our attention to some other ways in which differential equations may arise.

1.2.5 A two-dimensional curve

A curve is defined by the condition that the sum of the x and y intercepts of its normal is always equal to 2, Fig. 1.6. Express the condition by means of a differential equation.

The slope of the tangent at a point $P(x,y)$ on the curve is dy/dx. That of the normal is therefore $-1/(dy/dx)$. The equation of the normal is, with (X, Y) the coordinates of a general point on the line,

$$\frac{dy}{dx}(Y - y) = -(X - x)$$

The X intercept, OA, is $\quad x + y(dy/dx)$
and the Y intercept, OB, is $\quad y + x/(dy/dx)$.

The required condition is that $OA + OB = 2$, that is,

$$y + x + y\frac{dy}{dx} + x \bigg/ \frac{dy}{dx} = 2$$

or

$$y\left(\frac{dy}{dx}\right)^2 + (y + x - 2)\frac{dy}{dx} + x = 0 \qquad (1.16)$$

14 Ordinary Differential Equations

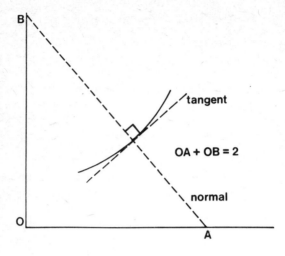

Fig. 1.6

1.2.6 Elimination of constants from equations

Example 1.2 Eliminate the constants a, b and c from the following:

(a) $y = x + a$

(b) $y = x^2 + a$

(c) $$y = \begin{cases} a & x < x_0 \\ b & x_0 \leqslant x < x_1 \\ c & x_1 \leqslant x \end{cases}$$

(d) $y = \sqrt{(a^2 - x^2)}$*

(e) $y = -\sqrt{(b^2 - x^2)}$

---

* By convention we take the value of the real number $\sqrt{x}$ to be positive. For example, $\sqrt{4} = 2$, while $4^{1/2} = \pm 2$.

(f) $x = \sin(\tau - a)$

(g) $y = ax + b$

(h) $x = a \cos \tau + b \sin \tau$

(i) $y = ae^{2x} + be^x$

In all cases we differentiate the equations so that a differential equation is obtained which does not involve the constants.

(a) $dy/dx = 1$.

(b) $dy/dx = 2x$.

(c) In this case note that the function is not differentiable at $x = x_0$ and $x = x_1$, and that the derivative is zero for all values of x other than $x = x_0$ and x_1.

Hence the differential equation obtained on eliminating the constants is

$$dy/dx = 0 \qquad x \neq x_0 \qquad x \neq x_1$$

(d)
$$\frac{dy}{dx} = -\frac{x}{\sqrt{(a^2 - x^2)}} \qquad x \neq \pm a$$

The function is not differentiable at $x = \pm a$, that is $y = 0$. Hence on replacing the square root by y

$$dy/dx = -x/y \qquad y > 0$$

(e) This is similar to the previous case, except that the function is defined for $y \leqslant 0$, hence on eliminating the constant b, we obtain

$$dy/dx = -x/y \qquad y < 0$$

(f) $dx/d\tau = \cos(\tau - a)$.

Hence since $\cos^2(\tau - a) = 1 - \sin^2(\tau - a)$ the differential equation obtained on eliminating a is

$$(dx/d\tau)^2 = 1 - x^2$$

We note that the same differential equation would result on eliminating a from $x = -\sin(\tau - a)$.

(g) In order to eliminate both a and b, the equation has to be differentiated twice, and hence

$$d^2y/dx^2 = 0$$

(h) Differentiating the equation twice

$$\frac{d^2x}{d\tau^2} = -(a\cos x + b\sin x)$$

Hence

$$\frac{d^2x}{d\tau^2} + x = 0$$

(i) We have

$$\frac{dy}{dx} = 2ae^{2x} + be^x \quad \text{and} \quad \frac{d^2y}{dx^2} = 4ae^{2x} + be^x$$

From these,

$$ae^{2x} = \frac{1}{2}\left(\frac{d^2y}{dx^2} - \frac{dx}{dy}\right) \quad \text{and} \quad be^x = 2\frac{dy}{dx} - \frac{d^2y}{dx^2}$$

Hence

$$y = \frac{1}{2}\left(\frac{d^2y}{dx^2} - \frac{dy}{dx}\right) + \left(2\frac{dy}{dx} - \frac{d^2y}{dx^2}\right)$$

or

$$\frac{d^2y}{dx^2} - 3\frac{dy}{dx} + 2y = 0$$

1.3 Classification of differential equations

The *order* of a differential equation is the order of the highest derivative that occurs.

In a differential equation involving y, x and the derivatives $d^n y/dx^n$, the variable x is called the *independent variable* and the variable y is called the *dependent variable*.

A differential equation is said to be *linear* if it is linear in the dependent variable and its derivatives; that is, if terms of the form y^2, $y(dy/dx)$, $(dy/dx)^2$, etc. do not appear in the differential equation.

The order and whether the equation is linear or non-linear are given for some differential equations below:

	order	linear
$\frac{dy}{dx} + y = x^2$	1	yes
$x^2 \frac{d^2y}{dx^2} + x\frac{dy}{dx} + y = 1$	2	yes

$$\frac{d^2y}{dx^2} + \left(\frac{dy}{dx}\right)^2 + y = x \quad 2 \qquad \text{no}$$

$$x\left(\frac{dy}{dx}\right) + \left(\frac{dy}{dx}\right)^2 - y = 0 \quad 1 \qquad \text{no}$$

The *degree* of each of the differential equations given above is the degree of the highest ordered derivative which occurs. The first three are of degree one and the fourth of degree two.

1.4 Solution of differential equations

We will consider firstly the solution of first-order differential equations with (in most cases) y as dependent variable and x as independent variable. Then as stated in Section 1.1 a function $f(x)$ is a solution of the differential equation if $y = f(x)$ satisfies the differential equation. The solution $y = f(x)$ describes a curve in the x,y plane and is called a *solution curve*. The angle θ that a tangent to the solution curve makes with the x axis is given by $\tan\theta = dy/dx$.

We now turn our attention to the first-order differential equations obtained in Section 1.2.6.

1.4.1 $dy/dx = 1$

For this differential equation, the solution curves are such that the tangents to the curves make the same angle to the x axis, namely $\theta = \pi/4$, regardless of its position. It is obvious, therefore, that a family of solution curves consists of parallel straight lines with $\tan\theta = 1$ (Fig. 1.7). Hence the equation of a family of solution curves is

$$y = x + a$$

It seems likely (but we cannot yet be certain) that the family of solution curves shown in Fig. 1.7 is the only family of curves with the property that the tangent to the curve has a constant angle to the x axis. It is also seems likely that there is a unique solution curve passing through any one point in the plane.

The function $x + a$ given by the solution curve $y = x + a$ is easily verified to be a solution to the differential equation, and it is noted

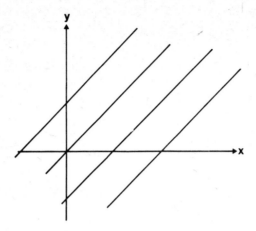

Fig. 1.7

that the solution $y = x + a$ is just the equation from which the differential equation was derived in Section 1.2.6(a).

Whatever other form of functions we take for y, we find that there is no other solution for the differential equation other than $y = x + a$. The solution $y = x + a$ is called the *general solution*. The general solution and the family of solution curves are equivalent. In physical situations, we shall be concerned with the solution of differential equations subject to given initial or boundary conditions. We might for instance in this case want the solution of the differential equation subject to $y = 1$ when $x = 0$. From the general solution we note that if we choose $a = 1$, then the condition is satisfied.

Since we have stated that there is no other form for the solution and since $a = 1$ is the only value of a such that the given condition is satisfied, then $y = x + 1$ is the unique solution of the differential equation subject to the given condition, and is said to be *the particular solution*.

The solution curve passing through the point $(0,1)$ is $y = x + 1$ and this solution curve is equivalent to the particular solution.

1.4.2 $dy/dx = 2x$

The slopes of the tangents to the solution curves of this differential equation are now dependent on x. In order to obtain an idea of the shape of the solution curves draw, at a grid of values of x,y, small elements whose slope to the x axis are given by $\tan^{-1} dy/dx$; that is, by $\tan^{-1} 2x$ (Fig. 1.8). These elements will be tangent to the solution curves.

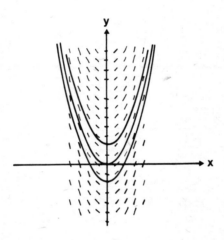

Fig. 1.8

The pattern of small elements gives an idea of the shape for the family of solution curves. Three such solution curves are sketched in Fig. 1.8. The curves sketched look like parabolas symmetric about the y axis, and it will be seen that the family of parabolas $y = x^2 + a$ are the required solution curves.

Again it appears that this family of parabolas is the only family of curves that have the desired tangents, and that there is a unique solution curve passing through any one point in the plane.

The function $x^2 + a$ can by direct substitution be shown to be a

20 Ordinary Differential Equations

solution of the differential equation, and $y = x^2 + a$ is the equation from which the differential equation was derived in Section 1.2.6(b).

Whatever other form of functions we take for y, there is no other form for the solution of the differential equation, and $y = x^2 + a$ is the general solution. For a given condition on y and x there is only one value for a, so that there is a unique solution of the equation subject to an initial condition.

1.4.3 $dy/dx = 0$, $x \neq x_0$ and x_1

Consider the construction of a solution curve starting at a point (X_0, Y_0), where $X_0 < x_0$ (Fig. 1.9).

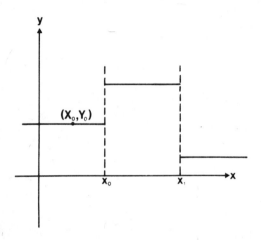

Fig. 1.9

For $x < x_0$ there is no problem in the construction of the solution curve, which is a straight line parallel to the x axis passing through the point (X_0, Y_0); that is, the line $y = Y_0$. However, since the slope of the solution curve is not defined for $x = x_0$, it is not possible to 'continue' the solution curve into the region $x > x_0$. In the region

$x_1 > x > x_0$ we note that $y = b$, where b is any constant, is a solution curve, and similarly in the region $x > x_1$, $y = c$ is a solution curve.

The general family of solution curves to the given differential equation is therefore

$$y = \begin{cases} a & x < x_0 \\ b & x_0 < x < x_1 \\ c & x_1 < x \end{cases}$$

where a is also an arbitrary constant.

There is not a unique solution curve passing through any given point (X_0, Y_0). However, in the neighbourhood of (X_0, Y_0) for which the derivative is defined, the solution curve is unique.

We note that the function defined above is a solution to the differential equation, and is the equation from which the differential equation was derived in Section 1.2.6(c).

1.4.4 $dy/dx = -x/y$

The slopes at a grid of points are shown in Fig. 1.10.

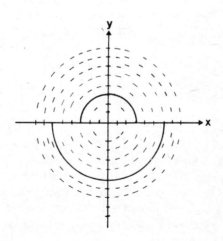

Fig. 1.10

It can be seen that solution curves are semicircles in the upper half plane and semicircles in the lower half plane. The solution curves are not circles, since the derivative is indeterminate at $y = 0$. The equations of the semicircles are given by

$$y = \sqrt{(a^2 - x^2)} \qquad y > 0$$
$$y = -\sqrt{(b^2 - x^2)} \qquad y < 0$$

It seems likely that there is a unique solution curve through any point in a region of the plane for which the derivative is everywhere defined. That is, the curve would then be restricted to either the upper or lower half planes. By direct substitution it can be seen that $y = \sqrt{(a^2 - x^2)}$ satisfies the differential equation, and since no other family of solutions in the upper half plane can be obtained it is the general solution. Similarly, $y = -\sqrt{(b^2 - x^2)}$ is the general solution in the lower half plane. For a given condition $y = y_0$ when $x = x_0$, there is a unique value of a^2 or b^2, that is there is a unique solution defined either in the upper or lower half plane.

1.4.5 $(dx/d\tau)^2 = 1 - x^2$

It will be noted that real values of $dx/d\tau$ exist only in the range $-1 \leqslant x \leqslant 1$, and that for any value of x in the range $-1 < x < 1$ there are two values for $dx/d\tau$. The derivatives at a grid of points are shown in Fig. 1.11.

It can be verified that solution curves are given by the harmonic sine or cosine curves, namely

$$x = \sin(\tau - a)$$

where a is any constant.

However, it can be seen from Fig. 1.11 that there are, for this differential equation, two special solution curves, $x = 1$ and $x = -1$, which are not members of the family of solutions $x = \sin(\tau - a)$. These special solutions are called *singular solutions*. It will be noted from Fig. 1.11 that the family of solution curves $x = \sin(\tau - a)$ form an *envelope*, the envelope being the singular solution curves $x = 1$ and $x = -1$.

Let us consider solution curves passing through the origin. The curves $x = \sin(\tau + n\pi)$, where n is any integer, are solution curves, but it can be seen that this family of curves reduces to two distinct curves

Introductory Ideas 23

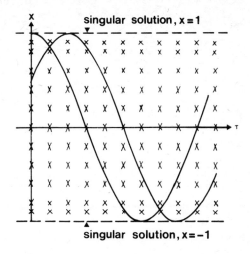

Fig. 1.11

$x = \pm \sin \tau$. However, the following curve is also a solution curve passing through the origin

$$x_1 = \begin{cases} \sin \tau & \tau \leqslant \pi/2 \\ 1 & \pi/2 \leqslant \tau \leqslant a + \pi/2 \\ \sin(\tau - a) & a + \pi/2 \leqslant \tau \end{cases}$$

as is also

$$x_2 = \begin{cases} \sin \tau & \tau \leqslant 3\pi/2 \\ -1 & 3\pi/2 \leqslant \tau \leqslant a + 3\pi/2 \\ \sin(\tau - a) & a + 3\pi/2 \leqslant \tau \end{cases}$$

There are therefore any number of solution curves through the origin, or indeed any point (τ_0, x_0) for $|x_0| \leqslant 1$. However, if we further state that we require the second derivative to be defined everywhere, then there are just two solutions namely, $x = \pm\sin \tau$. (For example, the family of curves x_1 given above has discontinuous second derivatives at $\tau = \pi/2$ and $\tau = a + \pi/2$.) There is a unique solution curve through any point (τ_0, x_0) ($|x_0| < 1$) if the sign of the derivative is also given at that point. It is to be noted that the value of the derivative cannot be specified as well as the value of x, but only its sign.

The solution $x = \sin(\tau - a)$ can by direct substitution be shown to be a solution of the differential equation, and since no other family of solutions exists it is the general solution. We note that $\sin(\tau - a + \pi) = -\sin(\tau - a)$ so that $x = -\sin(\tau - a)$ is a member of the general family of solutions and $x = \pm \sin(\tau - a)$ are just the equations from which the differential equation was derived in Section 1.2.6(f).

By direct substitution, $x = \pm 1$ can also be seen to be solutions which, as stated earlier, are the singular solutions.

1.4.6 $d^2y/dx^2 = 0$

A geometrical picture of the solution of a second-order differential equation is not as readily obtained as that of a first-order one, but a representation may be obtained by treating the second-order differential equation as a pair of first-order equations. This is done by letting dy/dx be a variable z.

Hence $d^2y/dx^2 = 0$ is replaced by the pair of first-order differential equations

$$dy/dx = z \qquad dz/dx = 0$$

where both y and z are functions of x. As x varies, the corresponding pairs of values (y, z) trace a curve in the y, z plane. This curve is a solution curve, since y and z satisfy the pair of first-order equations and hence y satisfies the original second-order equation. The y, z plane is often called the *phase space*.

From the pair of first-order equations we obtain

$$dz/dy = 0 \qquad \text{if } z \neq 0$$

Hence in the phase space, solution curves are given by $z = \text{const} = a$ (Fig. 1.12). It will be noted that for a given condition $y = y_0$, $dy/dx = z_0$ at $x = x_0$, that is for a given point (y_0, z_0) in the phase space, there is a unique solution through that point. This is true for any pair of values y_0, z_0.

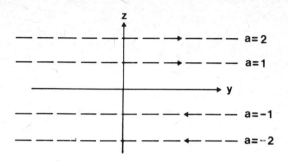

Fig. 1.12

It will be noted from $dy/dx = z$ that when $z > 0$, y increases as x increases and that when $z < 0$, y decreases as x increases. It follows therefore that solutions in the phase space move from left to right in the upper half plane and from right to left in the lower half plane, that is as x increases the solution proceeds in the direction of the arrows shown in Fig. 1.12.

The solution curves given in Fig. 1.12 give a relation between y and dy/dx. To obtain solution curves for y and x we can proceed in a similar manner to that already described in Section 1.4.1. That is, for each curve in the z plane a one-parameter family of curves is obtained in the x, y plane. Hence the complete set of solution curves in the x, y plane is a two-dimensional family.

We have seen that the differential equation $dy/dx = 1$ has a family of solution curves which are straight lines at an angle $\pi/4$ to the x axis, and with different intercepts on the y axis. That is, they are lines with equation $y = x + b$. Similarly, the solution curves to the differential equation $dy/dx = a$ are the family of straight lines given by $y = ax + b$, which are the set of all straight lines that can be drawn in the x, y plane.

For given conditions $y = y_0$, $z = z_0$ at $x = x_0$, we have seen that there is likely to be just one solution curve in the phase space, namely $z = z_0$, that is $dy/dx = z_0$. The resulting solution curves in the x, y plane are straight lines, and since the solution has to satisfy $y = y_0$ when $x = x_0$, it seems likely that there will be just a unique solution curve passing through that point, namely the line $y = z_0 (x - x_0) + y_0$.

By direct substitution, $y = ax + b$ can be shown to be a solution of the second-order differential equation $d^2y/dx^2 = 0$. No other form of solution can be obtained for this equation, and hence $y = ax + b$ is the general solution, and this is the equation from which the differential equation was derived in Section 1.2.6(g). To fix a and b, it is necessary to have two conditions given, and then they are determined uniquely. For instance, with $y = y_0$, $dy/dx = z_0$ at $x = x_0$, we set up two equations for a and b:

$$y_0 = ax_0 + b \qquad z_0 = a$$

Hence $a = z_0$, $b = y_0 - z_0 x_0$ and $y = z_0(x - x_0) + y_0$.

1.4.7 $d^2x/d\tau^2 + x = 0$

This is differential equation (1.13) derived in Section 1.2.2, and describes the motion of a mass attached to the end of a spring, but also describes many other physical situations.

With 'velocity' $dx/d\tau = y$, the second-order differential equation can be written as the pair of first-order equations

$$dx/d\tau = y \qquad dy/d\tau = -x$$

From these two equations,

$$dy/dx = -x/y \qquad y \neq 0$$

which is the first-order differential equation considered in Section 1.4.4.

Solution curves have been shown to be semicircles in the upper and lower half planes and with their centres at the origin. As x increases, the solution proceeds from left to right in the upper half plane and from right to left in the lower half plane; that is, the solution proceeds in a clockwise direction. However, for physical situations, we do not expect a jump in the value of x near $y = 0$ (that is, for zero velocity of the mass), so that the solution curves for this motion are circles centre the origin. (Difficulties have arisen in this case for values of y near zero, because we have divided by a quantity which tends to zero as y tends to zero.) Suppose now that we require the solution of the differential equation subject to the initial conditions $x = x_0$, $dx/d\tau = y_0$ when $\tau = 0$. From the construction of the solution, it seems likely that the unique solution curve in the phase space is a circle centre the origin and of radius $b = \sqrt{(x_0^2 + y_0^2)}$, that is

$$x^2 + y^2 = b^2$$

The solution curve in the phase space gives the important features of the motion described by the differential equation, namely:

(a) the 'displacement', x, has a maximum value b and a minimum value $-b$, and the mass moves back and forth between these values;
(b) the 'velocity', y has a maximum and minimum value;
(c) when the mass has its maximum or minimum 'displacement' it is at rest;
(d) when the mass has its maximum or minimum velocity the 'displacement' is zero.

This type of motion is known as simple harmonic motion.

The solution obtained relates displacement to velocity; we now proceed to obtain a solution which relates displacement to time.

Solution curves to the differential equation

$$(dx/d\tau)^2 = 1 - x^2$$

have been obtained in Section 1.4.5.

Solution curves to $(dx/d\tau)^2 = b^2 - x^2$ are obtained in a similar manner and are

$$x = b \sin(\tau - a)$$

with the singular solutions $x = \pm b$.

We see by direct substitution that the general solution satisfies the second-order differential equation (1.13) but that the singular solutions do not. (We note that $x = b$ implies that $dx/d\tau = y = 0$ which we have already seen needs care in its treatment.) Through any one point in the range $-b < x < b$, we have seen that there are two distinct solution curves, but when the derivative is also given there is a unique curve, that is there is a unique value for a.

The general solution can be written as

$$x = A \sin \tau + B \cos \tau$$

where $A = b \cos a$, $B = -b \sin a$, which again can be verified by direct substitution into (1.13). For a given initial condition $x = x_0$, $dx/d\tau = y_0$ when $\tau = 0$, values of A and B can be determined which will be unique.

1.4.8 The number of arbitrary constants in a solution

It will be noted from the above examples that in a region of the x, y plane for which dy/dx can be obtained, the general solution of a first-order differential equation results in a one-parameter family, that is, the general solution has one arbitrary constant. The general solution of a second-order differential equation is a two-parameter family, that is the general solution has two arbitrary constants. This can be generalised, and the solution of an nth-order differential equation results in an n-parameter family; that is, the general solution has n arbitrary constants.

1.4.9 Numerical solution

It is pertinent at this point to illustrate some of the underlying principles involved in the numerical solution of differential equations. We shall consider the differential equation

$$dy/dx = 1 + y^2$$

subject to $y = 0$ when $x = 0$, and suppose a solution is required for $x \geqslant 0$.

We study first a method due to Euler. Let the family of solution curves be as shown by dashed lines in Fig. 1.13, and the unique solution through $(0, 0)$ by the full curved line.

At $x = 0, y = 0$ the value of dy/dx as given by the differential equation is 1. For a small range of values of x near $x = 0$ the solution curve can be approximated by a straight line passing through the origin and with $dy/dx = 1$. The equation of this line is $y = x$. Through the point (x_1, x_1) passes a unique solution curve and the value of dy/dx for this curve can be obtained and is $1 + x_1^2$. The solution curve for the small range $x_1 < x < x_2$ is again approximated by a straight line whose equation is now given by

$$y = (1 + x_1^2)(x - x_1) + x_1$$

Our approximate value of y at $x = x_2$ is then given by

$$y_2 = (1 + x_1^2)(x_2 - x_1) + x_1$$

This process can be repeated and below are comparisons of the values of y correct to four decimal places from the solution curve and the approximate numerical values for 3 equal steps in x.

Introductory Ideas 29

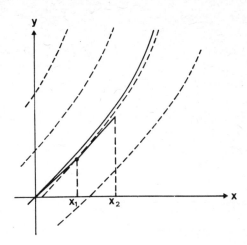

Fig. 1.13

x	0	0·1	0·2	0·3
y 'exact'	0	0·1003	0·2027	0·3093
y 'approximate'	0	0·1000	0·2010	0·3050

It is clear that there are various ways of improving our numerical solution. One way is to use the value of dy/dx at the 'mid point' of the solution curve rather than the initial point. This mid point is not known, but can be taken to be the value on the straight-line approximation already obtained. For our example, the mid point is approximated by the point (0·05, 0·05), and the value of dy/dx of the solution curve through and at that point is $1 + (0·05)^2 = 1·0025$. The straight-line approximation is therefore $y = 1·0025x$ and the value at $x = 0·1$ is 0·10025, which is a nearer approximation to the exact value. This process can be repeated to find the values at $x = 0·2$ and $x = 0·3$, which are 0·2028 and 0·3095 respectively.

We now consider Picard's method, which is based on the fact that solution curves for differential equations of the form $dy/dx = f(x)$ can be quite readily obtained. Geometrically the fact that the derivative is a function only of x eases the construction, while analytically the solution curves can be obtained by integration of $f(x)$ which will be discussed in Chapter 2.

We describe the method by considering the differential equation $dy/dx = 1 + y^2$, subject to $y = 0$ when $x = 0$. Our first approximate solution curve is obtained by taking y to be constant and equal to the given initial value. That is, we take $dy/dx = 1$. The solution curve for this differential equation and passing through the origin (see Section 1.4.1) is $y = x$. This is then taken as the second approximation, so that we take $dy/dx = 1 + x^2$. The solution curve for this which passes through the origin is

$$y = x + \tfrac{1}{3}x^3$$

The process is repeated to obtain y to the required accuracy, and the question of convergence is covered in Section 1.5.

1.5 Existence and uniqueness

It is not sufficient to say that because our differential equation represents a physical situation, a solution must exist and that this solution is unique. This argument could be sound if our mathematical model was a perfect representation of the physical situation, but we have seen that assumptions and approximations have been made in setting up a model. It is therefore of great importance to know whether a solution of the differential equation does exist, and if it does whether it is unique.

We have obtained general ideas of the existence and uniqueness of solutions of some differential equations in the previous section. In all cases, however, we assumed or stated that the solution curves drawn or the general solution obtained were the only possible families of solution. For simple differential equations, one can show that this is true, but for most equations it is not possible to do this directly. It is necessary therefore to use general theorems relating to the existence and uniqueness of solutions. The theory of existence and uniqueness is vast and many books are devoted entirely to this subject. Here we quote without proof some existence and uniqueness theorems.

We consider a first-order differential equation with a given condition on y and x, that is, an initial value problem of the form

$$dy/dx = f(x, y) \quad \text{with } y = y_0 \text{ when } x = x_0 \quad (1.17)$$

Existence theorem If $f(x, y)$ is continuous at all points (x, y) in some rectangle R given by

$$R: \quad |x - x_0| < a \qquad |y - y_0| < b$$

and bounded in R, that is,

$$|f(x, y)| \leq N \quad \text{for all } (x, y) \text{ in } R$$

where N is a constant, then the initial-value problem (1.17) has at least one solution $y(x)$ which is defined at least for all x in the interval $|x - x_0| < \alpha$, where α is the smaller of the two numbers a and b/N. It is to be noted that there may be a solution or solutions outside the range $|x - x_0| < \alpha$. The theorem states that provided the conditions are satisfied, there is certainly a solution in that range, it does not state that any solutions are confined to that range.

Uniqueness theorem If both $f(x, y)$ and $\partial f/\partial y$ are continuous for all (x, y) in the same rectangle R and bounded such that

(a) $|f| \leq N$ (b) $|\partial f/\partial y| \leq M$ for all (x, y) in R

where M and N are constants, then the initial-value problem (1.17) has only one solution $y(x)$ which is defined at least for all x in the interval $|x - x_0| < \alpha$, where α is as defined in the existence theorem. The solution can then be obtained by Picard's iteration method, the successive approximations converging to $y(x)$.

Example 1.3 $dy/dx = 1 + y^2$, with $y = 0$ when $x = 0$.

This is the initial value problem considered in Section 1.4.9. Take R as

$$|x| < 1 \qquad |y| < 1$$

that is, R is a square with its centre at the origin and with sides of length 2. Within this square the largest value of $|f(x, y)| \equiv |1 + y^2|$ is ≤ 2, hence $N = 2$. Now $\partial f/\partial y = 2y$, and hence within a square the largest value of $|\partial f/\partial y|$ is ≤ 2 so that $M = 2$.

Now $b/N = \frac{1}{2}$, and $a = 1$ so that $\alpha = \frac{1}{2}$. The existence theorem states that a solution or solutions exist in the range $-\frac{1}{2} < x < \frac{1}{2}$.

32 Ordinary Differential Equations

The uniqueness theorem states that there is a unique solution in this range.

We now quote an existence and uniqueness theorem for an initial-value problem for a second-order differential equation of the form

$$\frac{d^2y}{dx^2} + f(x)\frac{dy}{dx} + g(x)y = 0$$

with (1.18)

$$y = y_0 \text{ and } dy/dx = y'_0 \text{ when } x = x_0$$

Existence and uniqueness theorem If $f(x)$ and $g(x)$ are continuous functions in the range $\alpha < x < \beta$, and x_0 lies in that range, then the initial-value problem (1.18) has a unique solution $y(x)$ on $\alpha < x < \beta$.

Problem 1.1 Sketch solution curves for the following differential equations

(a) $dy/dx = y$ (b) $dy/dx = 1 + y^2$
(c) $dy/dx = 1/(1 + x^2)$ (d) $dy/dx = 1/x$
(e) $dy/dx = 2|x|$ (f) $dy/dx = (1 - y^2)/2$
(g) $dy/dx = y/x$

CHAPTER TWO

Analytic Solutions of First-Order Differential Equations

2.1 Introductory discussion

The procedure for solving differential equations discussed in Chapter 1 was basically to guess a solution by means of a geometrical construction, and then to verify that it indeed was a solution by substitution into the differential equation. However, there are many differential equations whose solutions can be obtained by the use of elementary integration. In this chapter, we consider first-order differential equations which can be solved in this manner. Integration is discussed in many books on calculus, and it is assumed that the student is familiar with the principles and techniques of this subject. Integration is facilitated by the use of tables of integrals and in this book reference will be made to particular formulae in Barnett and Cronin (1975).

2.2 Separable equations

The simplest form of first-order equation which can arise is

$$dy/dx = f(x) \qquad (2.1)$$

We need to find a function y whose derivative is identical to the given function f, and this is simply equivalent to the problem of integrating f. Hence y is the indefinite integral of f,

$$y(x) = \int f(x)\, dx + a \qquad (2.2)$$

where a is any constant.

With a given initial condition, $y = y_0$ when $x = x_0$, a unique value for a results and the solution is most simply written in terms of a definite integral

$$y(x) = y_0 + \int_{x_0}^{x} f(\eta)\, d\eta \qquad (2.3)$$

We assume that the function is integrable in the required domain.

Example 2.1 Solve $dy/dx = 2x$.

The solution is

$$y = \int 2x\, dx + a = x^2 + a$$

which is the general solution obtained in Section 1.4.2.

With the initial condition $y = 1$ when $x = 0$, the solution is

$$y = 1 + \int_0^x 2\eta\, d\eta = 1 + x^2$$

Uniqueness of the solution is given by the theorem of Section 1.5.

Another form of differential equation which can be solved by straightforward integration is

$$g(y)\frac{dy}{dx} = f(x) \qquad (2.4)$$

which is known as a *separable* equation. The solution is given by

$$\int g(y)\, dy = \int f(x)\, dx + a \qquad (2.5)$$

where a is any constant.

It is noted that (2.1) is a special case of (2.4) with $g(y) \equiv 1$.

With a given initial condition $y = y_0$ when $x = x_0$, the solution is

$$\int_{y_0}^{y} g(\xi)\, d\xi = \int_{x_0}^{x} f(\eta)\, d\eta \qquad (2.6)$$

Example 2.2 Cooling of a hot body.

The differential equation governing the cooling of a hot body has been obtained in Section 1.2.1 for Newton's law of cooling, and is

$$\frac{dT}{dt} = -k(T - T_m)$$

with $T = T_0$ when $t = 0$, $T_0 > T_m$.

This differential equation is easily rewritten in the form (2.4) with solution

$$\int_{T_0}^{T} \frac{d\xi}{\xi - T_m} = -k \int_0^t d\eta$$

With $x = \xi - T_m$, this becomes

$$\int_{T_0 - T_m}^{T - T_m} \frac{dx}{x} = -kt \quad \text{i.e.} \quad \ln \frac{T - T_m}{T_0 - T_m} = -kt$$

(see Barnett and Cronin (1975), Section 3.1.2). Hence

$$\frac{T - T_m}{T_0 - T_m} = e^{-kt}$$

or

$$T = T_m + (T_0 - T_m)e^{-kt}$$

Hence there is an exponential decay of the temperature from the initial temperature T_0 to that of the surrounding medium T_m.

Problem 2.1 Find the general solutions to the differential equations of Problem 1.1.

Problem 2.2 Find the solutions to the following differential equations with the given initial conditions:

(a) $2dx/dt = 1 - x^2$, with $x = 0$ when $t = 0$
(b) $v(a - x) dv/dx + ag = 0$, with $v = \sqrt{(2gh)}$ when $x = 0$
(c) $dy/dx = 2x\sqrt{(4 - y^2)}$, $x \geq 0$, with $y = 0$ when $x = \sqrt{(\pi 2)}$.

2.3 Homogeneous equations

Differential equations of the form

$$\frac{dy}{dx} = f\left(\frac{y}{x}\right) \qquad (2.7)$$

are called *homogeneous equations*. They can be transformed into separable equations by the substitution $y = tx$, where t is a function of x to be determined.

With $y = tx$ we have

$$dy/dx = t + x\, dt/dx$$

Hence (2.7) becomes

$$t + x\, dt/dx = f(t) \qquad (2.8)$$

or

$$\frac{1}{(f(t) - t)}\frac{dt}{dx} = \frac{1}{x}$$

which is of the form (2.4)

Example 2.3 In the chlorination of liquid benzene, the differential equation relating q, the mole fraction of benzene present, to r, the mole fraction of monochlorbenzene present, is

$$\frac{dr}{dq} = \frac{r}{8q} - 1 \qquad (2.9)$$

At the start of the reaction only benzene is present, so that $q = 1$ when $r = 0$. This is a homogeneous equation so we set $r = tq$. Then (2.8) gives, with x replaced by q,

$$t + q\, dt/dq = \tfrac{1}{8}t - 1$$

or

$$q\, dt/dq = -\tfrac{1}{8}(8 + 7t)$$

Hence

$$\int \frac{dq}{q} = -8\int \frac{dt}{8 + 7t} + \text{constant}$$

that is
$$\ln q = -\tfrac{8}{7} \ln (8 + 7t) + \text{constant}$$
(Barnett and Cronin (1975), Section 3.1.4) or
$$q = a(8 + 7t)^{-8/7}$$
Hence replacing t by r/q, and on using the initial condition
$$q = (1 + 7r/8q)^{-8/7}$$
or
$$r = \tfrac{8}{7}(q^{1/8} - q)$$

Problem 2.3 Find the general solutions to the following differential equations

(a) $xy \, dy/dx = x^2 + y^2$ (b) $y^2 + (3xy + x^2) \, dy/dx = 0$

2.4 Equations of the form $\dfrac{dy}{dx} = \dfrac{ax + by + c}{\alpha x + \beta y + \gamma}$

The variables x, y are transformed into the variables u, v by the transformation
$$x = u + m \qquad y = v + n$$
where m and n are constants which are to be determined.
Since m and n are constants,
$$\frac{dy}{dx} = \frac{dy}{dv}\frac{dv}{du}\frac{du}{dx} = \frac{dv}{du}$$
so that the differential equation becomes
$$\frac{dv}{du} = \frac{au + bv + (c + am + bn)}{\alpha u + \beta v + (\gamma + \alpha m + \beta n)}.$$
The constants m, n are now chosen (if possible) such that
$$c + am + bn = 0 \qquad \gamma + \alpha m + \beta n = 0$$

Then
$$\frac{dv}{du} = \frac{a + b(v/u)}{\alpha + \beta(v/u)}$$
which is of the form (2.7).

Example 2.4 Solve $\quad \dfrac{dy}{dx} = \dfrac{y - x + 1}{y + x + 5}$

Let
$$x = u + m \qquad y = v + n$$
with
$$1 - m + n = 0 \quad \text{and} \quad 5 + m + n = 0$$
Hence $m = -2$, $n = -3$, and
$$\frac{dv}{du} = \frac{v - u}{v + u}$$
Let $v = tu$; then
$$t + u\, dt/du = (t - 1)/(t + 1)$$
that is,
$$u\frac{dt}{du} = -\frac{1 + t^2}{1 + t}$$
or
$$\int \frac{1 + t}{1 + t^2} dt = -\int \frac{du}{u} + a$$

Hence $\quad \tan^{-1} t + \tfrac{1}{2} \ln(1 + t^2) = -\ln|u| + a$
(Barnett and Cronin (1975), Sections 3.1.2, 3.1.10), or
$$u^2 (1 + t^2) = b \exp[-2 \tan^{-1} t]$$
where b is a constant.

On substituting for u and t in terms of x and y, the solution is
$$(x + 2)^2 + (y + 3)^2 = b \exp\left[-2 \tan^{-1} \frac{y + 3}{x + 2}\right]$$

Analytic Solutions of First-Order Differential Equations

The method above fails if the differential equation can be written in the form

$$\frac{dy}{dx} = \frac{ax + by + c}{k(ax + by) + \gamma} \qquad \gamma \neq kc \qquad (2.10)$$

since the two equations for m, n are

$$am + bn + c = 0 \qquad k(am + bn) + \gamma = 0$$

which are inconsistent.

Note that when $\gamma = kc$, the differential equation (2.10) degenerates to $dy/dx = 1/k$.

To solve (2.10), make the substitution $z = ax + by$. Then

$$\frac{dz}{dx} = a + b\frac{dy}{dx} = a + b\left(\frac{z + c}{kz + \gamma}\right)$$

which is a separable equation.

Example 2.5 Solve

$$\frac{dy}{dx} = \frac{x + y + 1}{x + y + 2}$$

Let $z = x + y$, so that

$$\frac{dz}{dx} = 1 + \frac{dy}{dx}$$

That is

$$\frac{dz}{dx} = 1 + \frac{z + 1}{z + 2} = \frac{2z + 3}{z + 2}$$

Hence

$$\int \frac{z + 2}{2z + 3} \, dz = \int dx + a$$

that is

$$\tfrac{1}{2} z + \tfrac{1}{4} \ln |2z + 3| = x + a$$

or, after rearrangement,

$$2x + 2y + 3 = b \exp 2(x - y)$$

40   Ordinary Differential Equations

Problem 2.4   Find the general solutions to the following differential equations:

(a) $\dfrac{dy}{dx} = \dfrac{x - y + 1}{x - 2y + 3}$

(b) $\dfrac{dy}{dx} = \dfrac{2x - 5y + 3}{2x + 4y - 6}$

(c) $\dfrac{dy}{dx} = \dfrac{x - y - 1}{x - y - 5}$

2.5   Exact differential equations

On differentiating the equation $f(x, y) =$ constant with respect to x, the following differential equation results:

$$\frac{\partial f}{\partial x} + \frac{\partial f}{\partial y}\frac{dy}{dx} = 0 \tag{2.11}$$

We note that for functions f such that $\partial f/\partial x$, $\partial f/\partial y$, $\partial^2 f/\partial x \partial y$ and $\partial^2 f/\partial y \partial x$ exist and are continuous, then

$$\frac{\partial^2 f}{\partial y \partial x} = \frac{\partial^2 f}{\partial x \partial y}$$

We also note that the solution of (2.11), which is called an *exact differential equation*, is $f(x, y) =$ constant.

Consider now a differential equation

$$g(x, y) + h(x, y)\frac{dy}{dx} = 0 \tag{2.12}$$

Comparing (2.11) and (2.12), the latter is an exact differential equation provided that g and h can be written as $\partial f/\partial x$ and $\partial f/\partial y$ respectively, where f is a function of x and y to be determined. If g and h can be written in this manner, then $\partial g/\partial y = \partial h/\partial x$. This is the condition that equation (2.12) be exact, and if it is exact, then the solution is $f(x, y) =$ constant, where $g = \partial f/\partial x$ and $h = \partial f/\partial y$.

Example 2.6   A spring of negligible weight hangs vertically with the upper end fixed and with a mass of m kg attached to the lower end. If the mass is moving with velocity v_0 m/s when the spring is unstretched, find the velocity v as a function of the extension x m.

Analytic Solutions of First-Order Differential Equations

This is the system discussed in Section 1.2.2. The motion of the mass is given by (1.11), where m_1 is to be replaced by m and where z_1 is the length of the spring. Hence in terms of the extension $x = z_1 - L$, the differential equation describing the motion is

$$m \frac{d^2 x}{dt^2} + \frac{\lambda x}{L} = mg$$

Since

$$\frac{d^2 x}{dt^2} = \frac{dv}{dt} = \frac{dx}{dt}\frac{dv}{dx} = v\frac{dv}{dx}$$

where v is velocity, we have

$$mg - \frac{\lambda}{L} x - mv \frac{dv}{dx} = 0 \qquad (2.13)$$

Now

$$\frac{\partial}{\partial v}\left(mg - \frac{\lambda}{L} x\right) = 0 \qquad \text{and} \qquad \frac{\partial}{\partial x}(-mv) = 0$$

hence (2.13) is exact with solution $f(v, x) =$ constant, where

$$\frac{\partial f}{\partial x} = mg - \frac{\lambda}{L} x \qquad \text{and} \qquad \frac{\partial f}{\partial v} = -mv \qquad (2.14\text{a, b})$$

On integrating (2.14a) with respect to x, we obtain

$$f(v, x) = mgx - \frac{\lambda}{2L} x^2 + V(v) \qquad (2.15)$$

where V, a function of v, replaces the usual constant of integration, since integration is performed with v regarded as constant.

On integrating (2.14b) with respect to v we obtain similarly

$$f(v, x) = X(x) - \tfrac{1}{2} mv^2 \qquad (2.16)$$

From (2.15) and (2.16), it follows that

$$V(v) = -\tfrac{1}{2} mv^2 \qquad \text{and} \qquad X(x) = mgx - \lambda x^2/2L$$

Hence the general solution of (2.13) is

$$mgx - \frac{\lambda x^2}{2L} - \frac{mv^2}{2} = a$$

42   Ordinary Differential Equations

With the given initial condition $v = v_0$ when $x = 0$, it follows that $a = -\frac{1}{2}mv_0^2$, so that

$$x\left(mg - \frac{\lambda x}{2L}\right) = \frac{1}{2}m(v^2 - v_0^2)$$

The solution of (2.13) could more easily have been obtained by treating it as of separable form, that is

$$\int_0^x \left(mg - \frac{\lambda}{L}\xi\right) d\xi = m \int_{v_0}^v \eta \, d\eta$$

which leads to the same solution as already obtained.

This illustrates that some differential equations can be solved analytically in more than one way (although the solutions obtained can always be written in the same identical form).

Example 2.7   Solve

$$(4x^3 y^2 + \sin x) + (2x^4 y + \cos y)\frac{dy}{dx} = 0$$

Clearly

$$\frac{\partial}{\partial y}(4x^3 y^2 + \sin x) = 8x^3 y = \frac{\partial}{\partial x}(2x^4 y + \cos y)$$

so the differential equation is exact with solution $f(x, y) = a$, where

$$\frac{\partial f}{\partial x} = 4x^3 y^2 + \sin x \quad \text{and} \quad \frac{\partial f}{\partial y} = 2x^4 y + \cos y$$

From these

$$f = x^4 y^2 - \cos x + Y(y) \quad \text{and} \quad f = x^4 y^2 + \sin y + X(x)$$

At this stage a partial check is obtained since terms involving both x and y must be common to the two expressions, in the above $x^4 y^2$ being the common term. Hence $X(x) = -\cos x$ and $Y(y) = \sin y$, and the required solution is

$$x^4 y^2 - \cos x + \sin y = a$$

Problem 2.5   Find the general solutions to the following differential equations:

(a) $x \sec^2 y \, dy/dx + \tan y - e^x = 0$
(b) $(x^2 \cos y + \tan x + 2y) dy/dx + 2x \sin y + (1 + y) \sec^2 x = 0$

2.6 Linear equations

We consider the linear equation

$$\frac{dy}{dx} + P(x)y = Q(x) \qquad (2.17)$$

where P and Q are given functions of x. This equation is not exact, since

$$\frac{\partial}{\partial y}[P(x)y - Q(x)] = P(x) \quad \text{and} \quad \frac{\partial}{\partial x}(1) = 0$$

The equation can, however, be made exact on multiplying both sides of the equation by an *integrating factor*, $R(x)$. That is, we consider

$$R(x)P(x)y - R(x)Q(x) + R(x)\frac{dy}{dx} = 0 \qquad (2.18)$$

This is exact when

$$R(x)P(x) = dR/dx \qquad (2.19)$$

The differential equation (2.19) is of separable form, with solution

$$R(x) = a \exp\left[\int P(x)\,dx\right]$$

Since we are looking for any function $R(x)$ which will make the equation exact, the constant a can be chosen to have any value, and will for convenience be taken as unity. The solution of (2.18) is $f(x, y) = $ constant, where

$$\frac{\partial f}{\partial x} = R(x)P(x)y - R(x)Q(x) \quad \text{and} \quad \frac{\partial f}{\partial y} = R(x)$$
$$(2.20\text{a, b})$$

Now since, from (2.19),

$$\int R(x)P(x)\,dx = \int \frac{dR}{dx}\,dx = R(x)$$

it follows by integrating (2.20a) with respect to x that

$$f = R(x)y - \int R(x)Q(x)\,dx + Y(y)$$

Ordinary Differential Equations

and from (2.20b)

$$f = R(x)y + X(x)$$

This gives

$$X(x) = -\int R(x)Q(x)\,dx \quad \text{and} \quad Y(y) = 0$$

so that the solution is $f(x, y)$ = constant, that is,

$$R(x)y - \int R(x)Q(x)\,dx = b$$

or

$$y = \frac{1}{R(x)}\left[b + \int R(x)Q(x)\,dx\right] \tag{2.21}$$

where

$$R(x) = \exp\left[\int P(x)\,dx\right]$$

and b is an arbitrary constant.

Example 2.8 Solve $y' + y \tan x = \cos x$.
The integrating factor is given by

$$R(x) = \exp\left[\int \tan x\,dx\right] = \exp\,[\ln \sec x] = \sec x$$

(Barnett and Cronin (1971), Section 3.3.3). Hence the solution is given by

$$y = \frac{1}{\sec x}\left[b + \int \sec x \cos x\,dx\right]$$

that is,

$$y = (b + x) \cos x$$

Example 2.9 The current i (A) in a simple series electric circuit consisting of a battery providing a constant electromotive force E_0 (V), an inductance of L (H) and resistance R (ohms) is given by

$$L\,di/dt + Ri = E_0 \tag{2.22}$$

Find the current at time t, when initially $i = i_0$.

Analytic Solutions of First-Order Differential Equations   45

We have that $P(t) = R/L$ and $Q(t) = E_0/L$. The integrating factor is
$$R(t) = \exp\left[\int (R/L)\, dt\right] = \exp\left[(R/L)t\right]$$
Hence the solution is
$$i(t) = \exp\left[-Rt/L\right]\left\{b + \int (E_0/L) \exp(Rt/L)\, dt\right\}$$
$$= \exp\left[-Rt/L\right]\{b + (E_0/R) \exp(Rt/L)\} = E_0/R + b \exp(-Rt/L)$$
Using the given initial condition, we obtain $b = i_0 - E_0/R$; hence
$$i(t) = i_0 \exp(-Rt/L) + (E_0/R)[1 - \exp(-Rt/L)]$$

As time becomes large, the current tends to the constant value E_0/R and this is called the *steady-state* solution of (2.22). We note that this steady-state solution is independent of the initial condition, and that it could have been obtained without solution of the differential equation by considering the sign of the derivative as a function of i (Fig. 2.1).

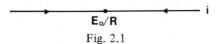

E_0/R

Fig. 2.1

The current decreases ($di/dt < 0$) when $i > E_0/R$, and increases when $i < E_0/R$, and since $di/dt = 0$ when $i = E_0/R$, the current must eventually tend to the value E_0/R.

Example 2.10 Solve
$$\frac{dr}{dq} = \frac{r}{8q} - 1 \quad \text{with } r = 0 \text{ when } q = 1$$

This is (2.9), already solved in Example 2.3, but we now solve it using the integrating factor
$$R(x) = \exp\left[\int -\frac{dq}{8q}\right] = \exp\left[-\tfrac{1}{8} \ln q\right] = q^{-1/8}$$
The solution is therefore
$$r = q^{1/8}\left[a + \int q^{-1/8}(-1)\, dq\right] = q^{1/8}[a - \tfrac{8}{7} q^{7/8}]$$
(Barnett and Cronin (1975), Section 3.1.3)

Using the given initial condition, we obtain $a = \frac{8}{7}$, hence

$$r = \tfrac{8}{7}[q^{1/8} - q]$$

This agrees with the solution already given in Example 2.3.

Problem 2.6 Find the general solution of the following differential equations:

(a) $dy/dx + 2y/x = 4x$ \qquad (b) $(1 + x^2)\,dy/dx + xy = 1$
(c) $\sin x\, dy/dx - y \cos x = 2 \sin^3 x$.

2.7 Bernoulli's equation

The differential equation of the form

$$dy/dx + P(x)y = Q(x)y^n \qquad n \neq 0, 1 \qquad (2.23)$$

is known as *Bernoulli's equation*.

This equation can be transformed into linear form by means of the substitution $z = y^{1-n}$. Multiplying both sides of (2.33) by y^{-n} and using $dz/dx = (1-n)y^{-n}\,dy/dx$ leads to

$$\frac{1}{(1-n)}\frac{dz}{dx} + P(x)y^{1-n} = Q(x)$$

that is,

$$dz/dx + (1-n)P(x)z = (1-n)Q(x)$$

which is of the form of (2.17).

Example 2.11 The differential equation governing the motion of a mass attached to the end of a spring subject to a damping force proportional to the square of the velocity is

$$v\,dv/dx - kv^2 + \lambda x = 0$$

where v is velocity and x is the extension of the spring. The spring is stretched a distance x_0 and released from rest. Determine the position at which the spring is instantaneously at rest for the first time after its release.

The differential equation can be rewritten as

$$dv/dx - kv = -\lambda x/v$$

and is therefore a Bernoulli equation with $n = -1$, $P(x) = -k$ and $Q(x) = -\lambda x$. Hence let $z = v^2$, so that

$$dz/dx - 2kz = -2\lambda x$$

This is a linear equation with integrating factor

$$R(x) = \exp\left[\int -2k\,dx\right] = \exp(-2kx)$$

and solution given by (2.21):

$$z = \exp(2kx)\left[b + \int(-2\lambda x)\exp(-2kx)\,dx\right]$$
$$= b\exp(2kx) + (\lambda/2k^2)(1 + 2kx)$$

(Barnett and Cronin (1975), Section 3.4.3).

With $z = v^2$, and using the initial condition $v = 0$ when $x = x_0$, we obtain

$$v^2 = (\lambda/2k^2)\{1 + 2kx - (1 + 2kx_0)\exp[-2k(x_0 - x)]\}$$

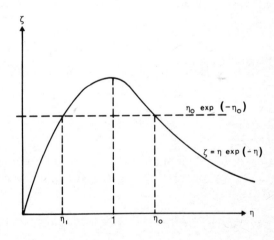

Fig. 2.2

48   Ordinary Differential Equations

Writing $1 + 2kx = \eta$ and $1 + 2kx_0 = \eta_0$, the velocity is zero if
$$\eta = \eta_0 \exp[-(\eta_0 - \eta)]$$
or
$$\eta \exp(-\eta) = \eta_0 \exp(-\eta_0)$$

This equation has to be solved numerically, and gives two values for η (see Fig. 2.2). One of these is the initial condition, and the other gives the required position of the spring, which is $x = -(1 - \eta_1)/2k$. (Notice that x and η decrease from their initial values x_0 and η_0.)

Problem 2.7   Find the solution of the differential equation
$$dy/dx = xy^3 - y$$
with the initial condition $y = 1$ when $x = 0$.

2.8   Miscellaneous examples

Some first-order differential equations may be solved by replacing the derivative dy/dx by p and differentiating the equation either with respect to x or with respect to y. We consider some cases.

2.8.1   Equations of the form $y = f(x, p)$ (2.24)

Differentiate with respect to x: then
$$dy/dx = p = \partial f/\partial x + (\partial f/\partial p)(dp/dx)$$

This is a first-order differential equation with variables p and x which can be solved by methods already described, and with general solution, say
$$\phi(x, p, c) = 0 \qquad (2.25)$$
where c is the arbitrary constant.

The solution of (2.24) is then obtained by eliminating p between (2.24) and (2.25), or by treating the pair of equations as the parametric equations for y and x.

Example 2.12   Solve $x(dy/dx)^2 - 2y(dy/dx) + 4x = 0$.

The equation can be rewritten as
$$y = \tfrac{1}{2}xp + 2x/p$$

Differentiating with respect to x, we obtain
$$p = (\tfrac{1}{2}p + 2/p) + (\tfrac{1}{2}x - 2x/p^2)\,dp/dx$$
Hence, on simplifying,
$$dp/dx = p/x$$
which is a separable equation with solution
$$p = ax$$
This can be substituted back into the differential equation to give the required solution
$$y = \tfrac{1}{2}ax^2 + 2/a$$

Example 2.13 $\quad y = px + f(p)$

This is a special form of eq. (2.24), and is called a *Clairaut equation*. Differentiate with respect to x:
$$p = p + (x + df/dp)\,dp/dx$$
Hence either $(x + df/dp) = 0$ or $dp/dx = 0$.
With $dp/dx = 0$, that is with $p = a$, the general solution of the Clairaut equation is
$$y = ax + f(a) \qquad (2.26)$$
We note that the general solution is simply obtained by replacing the derivative p in the differential equation by the arbitrary constant a.
With $x = -df/dp$, we obtain
$$y = -p\,df/dp + f(p)$$
This is a parametric equation for x and y which does not contain an arbitrary constant and is a *singular* solution.

Example 2.14 $\quad y = px + \sqrt{(4 + p^2)}$

The general solution is given by (2.26) as
$$y = ax + \sqrt{(4 + a^2)}$$
This is a family of straight lines (Fig. 2.3). The singular solution is given by
$$x = -df/dp = -p/\sqrt{(4 + p^2)}$$
$$y = -p^2/\sqrt{(4 + p^2)} + \sqrt{(4 + p^2)} = 4/\sqrt{(4 + p^2)}$$

50   Ordinary Differential Equations

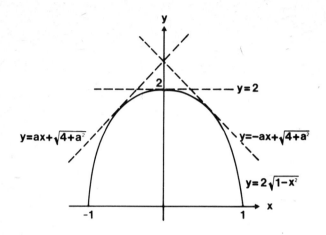

Fig. 2.3

On eliminating p, we have
$$y = 2\sqrt{(1-x^2)}$$
This is an equation for an ellipse, to which the family of straight lines are tangent; that is, the ellipse is the *envelope* of the family of straight lines.

2.8.2   **Equations of the form** $x = f(y, p)$ (2.27)

Differentiate with respect to y: then
$$dx/dy = 1/p = \partial f/\partial y + (\partial f/\partial p)(dp/dy)$$
This equation can then be solved by methods already described.

Example 2.14   Determine the shape of a reflecting mirror such that light coming from a fixed source is reflected in parallel rays.

We take the source to be at the origin and the parallel rays to be in the direction of the z axis, as in Fig. 2.4. From symmetry considerations the reflector is a surface of revolution with its axis along the z axis. Hence, if we use cylindrical polar coordinates r, θ, z, the equation of the surface is independent of θ.

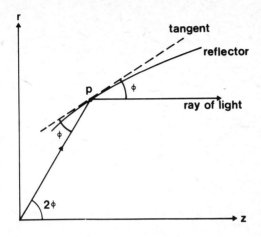

Fig. 2.4

If we use the condition that the angle of incidence of a light ray with respect to the tangent to the reflector at a point P is equal to the angle of reflection, the following equation results:

$$\frac{r}{z} = \tan 2\phi = \frac{2 \tan \phi}{1 - \tan^2 \phi} = \frac{2\, dr/dz}{1 - (dr/dz)^2} = \frac{2p}{1 - p^2}$$

Hence

$$2z = r/p - rp$$

which has the form (2.27).

Differentiate with respect to r:

$$2/p = 1/p - p - (r/p^2 + r)\, dp/dr.$$

Hence

$$dp/dr = -p/r$$

with solution obtained by separation of variables as

$$p = a/r$$

Hence the shape of the reflector is
$$2z = r^2/a - a$$
or
$$r^2 = 2az + a^2$$
which is a member of the family of paraboloids of revolution with focus the origin.

Problem 2.8 Find the general solutions of the following differential equations:

(a) $y = (1 + dy/dx)x + 2(dy/dx)^2$
(b) $(dy/dx)^2 = y - x(dy/dx)$
(c) $2y = (dy/dx)[6x + 3y^2(dy/dx)]$

Problem 2.9 Find and sketch the family of general solutions and singular solutions of the differential equation
$$x(dy/dx)^2 - 2y(dy/dx) + x = 0$$

Miscellaneous problems

Problem 2.10 Find the general solutions of the following differential equations:

(a) $(1 + x^2)dy/dx + xy = ky$
(b) $dy/dx = xe^{2x+y}$
(c) $x\,dy/dx = y + x\exp(-y/x)$
(d) $\dfrac{dy}{dx} = \dfrac{x + 2y - 1}{2x + 4y + 3}$
(e) $x \sin y\, dy/dx = \cos y + x^2 e^x + 2xe^x$
(f) $x^2 \sin x\, dy/dx + (x^2 \cos x - x \sin x)y = 1 + x$
(g) $x\, dy/dx + (2x^2 + 1)y = \sin 5x \exp(-x^2)$
(h) $dy/dx + y = 2y^2(\cos x - \sin x)$
(i) $x\, dy/dx - y = (dy/dx)^2$
(j) $(dy/dx)^3 + 8y^2 = 4xy(dy/dx)$

Problem 2.11 Find the solutions to the following differential equations subject to the given conditions:

(a) $(x + 3y) \, dy/dx = x - y$, with $dy/dx = 0$ when $x = 1$
(b) $x \, dy/dx - y = (x + 1)^2$, with $y = 0$ when $x = 1$
(c) $dy/dx - y = 4xy^5$, with $y \to \infty$ when $x \to 0$

CHAPTER THREE

Numerical Solutions of First- Order Differential Equations

3.1 Introductory discussion

In Chapter 2, analytic solutions of first-order differential equations were obtained, but as stated in the introduction, many differential equations are difficult to solve analytically. In such cases it is necessary to turn to numerical methods of solution. Various methods of solution are developed and discussed in this chapter, starting with the solution of separable differential equations which can be obtained by integration.

3.2 Integration formulae

We consider solutions of the differential equation $dy/dx = f(x)$, subject to the initial condition $y = y_0$ when $x = x_0$.

The solution of this differential equation given by (2.3) is

$$y(x) = y_0 + \int_{x_0}^{x} f(\eta)\, d\eta$$

The numerical solution of the differential equation (and hence the numerical value for the integral of f) will be obtained using the idea introduced in Section 1.4.9; that is, the solution y will be approximated by the value of y obtained from straight-line elements.

Let the family of solution curves be as shown in Fig. 3.1, where we note that because the slope of the solution curve is independent of y, the curves are parallel.

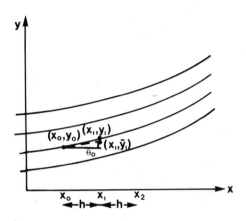

Fig. 3.1

The particular solution curve required passes through the point (x_0, y_0). We first obtain the approximate solution $\bar{y}_1$ at $x = x_1$, where $x_1 - x_0 = h$. This approximate value $\bar{y}_1$ is obtained from a straight line passing through the point (x_0, y_0) and making an angle θ_0 to the x axis (see Fig. 3.1). Hence

$$\bar{y}_1 = y_0 + h \tan \theta_0$$

The value of $\tan \theta_0$ can be taken to be the value of the slope of the solution curve at any point in the range $x_0 \leqslant x \leqslant x_1$ or any weighted average of the slopes in the same range. Four of the infinite number of values that may be taken for $\tan \theta_0$ are

$$f_{1/2} \quad \tfrac{1}{2}(f_0 + f_1) \quad \tfrac{1}{6}(f_0 + 4f_{1/2} + f_1)$$
$$\tfrac{1}{20}(f_0 + 5f_{1/6} + f_{2/6} + 6f_{3/6} + f_{4/6} + 5f_{5/6} + f_1)$$

where $\tan \theta = dy/dx = f(x)$ and where we take $f_r = f(x_r)$, with $x_r = x_0 + rh$.

If the solution curve is known (which of course it is not), then a value of $\tan \theta_0$ can be found such that $y_1 = \bar{y}_1$. It can be verified, for example, that if the solution curve is given by a quadratic in x, then $\tan \theta_0$ can be taken as $\frac{1}{2}(f_0 + f_1)$, and if the solution curve is given by a cubic, then $\tan \theta_0$ can be taken as $\frac{1}{6}(f_0 + 4f_{1/2} + f_1)$.

Example 3.1 Verify that if $y = ax^2 + bx + c$, $x_0 \leq x \leq x_1$, then with $\tan \theta_0 = \frac{1}{2}(f_0 + f_1)$, $y_1 = \bar{y}_1$.

We have

$$y_0 = ax_0^2 + bx_0 + c \qquad y_1 = ax_1^2 + bx_1 + c$$

$$h = x_1 - x_0 \qquad f_0 = (dy/dx)_{x=0} = 2ax_0 + b \qquad f_1 = 2ax_1 + b$$

Hence

$$\bar{y}_1 = y_0 + \tfrac{1}{2}h(f_0 + f_1)$$
$$= ax_0^2 + bx_0 + c + \tfrac{1}{2}(x_1 - x_0)[2a(x_0 + x_1) + 2b]$$
$$= ax_0^2 + bx_0 + c + a(x_1^2 - x_0^2) + b(x_1 - x_0)$$
$$= ax_1^2 + bx_1 + c$$
$$= y_1$$

In practice the appropriate value of $\tan \theta_0$ is obtained by assuming that over the step x_0 to x_1 the solution curve is approximated by a known curve. In most cases the step size is taken small enough so that the solution curve is approximated by a quadratic or cubic.

The value of the solution at $x = x_2$ after another step of length h can be approximated in a similar manner so that

$$\bar{y}_2 = \bar{y}_1 + h \tan \theta_1 = y_0 + h(\tan \theta_0 + \tan \theta_1)$$

In general after n steps

$$\bar{y}_n = y_0 + h(\tan \theta_0 + \tan \theta_1 + \cdots + \tan \theta_{n-1})$$

We now take the four weighted averages already listed, which lead to different named formulae for numerical integration.

(a) *Mid point rule*. We take $\tan \theta_r = f_{r+1/2}$; hence
$$y_n \simeq \bar{y}_n = y_0 + h[f_{1/2} + f_{3/2} + \cdots + f_{n-\frac{1}{2}}]$$
and
$$\int_{x_0}^{x_n} f(\eta)\, d\eta \simeq h[f_{1/2} + f_{3/2} + \cdots + f_{n-\frac{1}{2}}] \qquad (3.1)$$

(b) *Trapezoidal rule*. We take $\tan \theta_r = \tfrac{1}{2}(f_r + f_{r+1})$; hence
$$y_n \simeq \bar{y}_n = y_0 + h[\tfrac{1}{2}(f_0 + f_1) + \tfrac{1}{2}(f_1 + f_2) + \cdots + \tfrac{1}{2}(f_{n-1} + f_n)]$$
$$= y_0 + h[\tfrac{1}{2}f_0 + f_1 + f_2 + \cdots + f_{n-1} + \tfrac{1}{2}f_n]$$
and
$$\int_{x_0}^{x_n} f(\eta)\, d\eta \simeq h[\tfrac{1}{2}f_0 + f_1 + f_2 + \cdots + f_{n-1} + \tfrac{1}{2}f_n] \qquad (3.2)$$

(c) *Simpson's rule*. We take $\tan \theta_r = \tfrac{1}{6}(f_r + 4f_{r+1/2} + f_{r+1})$; hence
$$y_n \simeq \bar{y}_n = y_0 + \tfrac{1}{6}h[(f_0 + 4f_{1/2} + f_1) + (f_1 + 4f_{3/2} + f_2) + \cdots$$
$$+ (f_{n-1} + 4f_{n-1/2} + f_n)]$$
$$= y_0 + \tfrac{1}{6}h[f_0 + 4f_{1/2} + 2f_1 + 4f_{3/2} + 2f_2 + \cdots$$
$$+ 2f_{n-1} + 4f_{n-1/2} + f_n]$$

We see that the derivatives have to be evaluated at the mid-points as well as the end points of the intervals. Hence if we relabel so that
$$x_0 \to x_0, x_{1/2} \to x_1, x_1 \to x_2, \cdots, x_n \to x_m \quad (m \text{ now even})$$
and let $h = 2H$, so that $x_m - x_0 = mH$, then
$$\int_{x_0}^{x_m} f(\eta)\, d\eta \simeq \tfrac{1}{3}H[f_0 + 4f_1 + 2f_2 + 4f_3 + \cdots + 2f_{m-2} + 4f_{m-1} + f_m] \qquad (3.3)$$

It has already been stated that this rule gives an exact value for y when the solution curve is a cubic. This is equivalent to stating, since $f(x)$ is the derivative of $y(x)$, that Simpson's rule gives the exact value for the integral if the integrand is a quadratic. It should also be noted

58 Ordinary Differential Equations

at this point that although a weighting of three derivatives would normally only be expected to give the exact value of y for a cubic solution curve, Simpson's rule does in fact, freakishly, give the exact value for a quartic solution curve. That is, Simpson's rule also gives the exact value for the integral if the integrand is a cubic.

Example 3.2 Find numerically the solution at $x = 1$ of the differential equation $(1 + x^2)\,dy/dx = 1$ with $y = 0$ when $x = 0$.

The value of the solution is $y(1) = \int_0^1 (1 + x^2)^{-1}\,dx$, so that the solution gives a numerical value for the definite integral. We note that in this case the integral can be evaluated analytically and is $[\tan^{-1} x]_0^1 = \pi/4 = 0\cdot785\ 398$, to six decimal places.

The values of f_r with an interval of length $\tfrac{1}{8}$ are

f_0	f_1	f_2	f_3	f_4
1·000 000	0·984 615	0·941 176	0·876 712	0·800 000

f_5	f_6	f_7	f_8
0·719 101	0·640 000	0·566 372	0·500 000

(a) Trapezoidal approximation
  (i) 2 intervals, $h = \tfrac{1}{2}$.
   $\bar{y}(1) = \tfrac{1}{2}\,[0\cdot500\ 000 + 0\cdot800\ 000 + 0\cdot250\ 000] = 0\cdot775\ 000$
  (ii) 4 intervals, $h = \tfrac{1}{4}$.
   $\bar{y}(1) = \tfrac{1}{4}\,[0\cdot500\ 000 + 0\cdot941\ 176 + 0\cdot800\ 000 + 0\cdot640\ 000 + 0\cdot250\ 000] = 0\cdot782\ 794$
  (iii) 8 intervals, $h = \tfrac{1}{8}$.
   $\bar{y}(1) = \tfrac{1}{8}[\tfrac{1}{2}f_0 + f_1 + f_2 + f_3 + f_4 + f_5 + f_6 + f_7 + \tfrac{1}{2}f_8] = 0\cdot784\ 747$
(b) Simpson's approximation
  (i) 2 intervals, $H = \tfrac{1}{2}$.
   $\bar{y}(1) = [1/(2\cdot3)]\,[1\cdot000\ 000 + 4(0\cdot800\ 000) + 0\cdot500\ 000]$
   $= 0\cdot783\ 333$
  (ii) 4 intervals, $H = \tfrac{1}{4}$.
   $\bar{y}(1) = [1/(4\cdot3)]\,[f_0 + 4f_2 + 2f_4 + 4f_6 + f_8] = 0\cdot785\ 392$
  (iii) 8 intervals, $H = \tfrac{1}{8}$.
   $\bar{y}(1) = [1/(8\cdot3)]\,[f_0 + 4f_1 + 2f_2 + 4f_3 + 2f_4 + 4f_5 + 2f_6 + 4f_7 + f_8]$
   $= 0\cdot785\ 398$

Let us suppose that the solution is required to four decimal places. How do we know that the values given by the approximations are correct to the required accuracy?

Numerical Solutions of First-Order Differential Equations 59

One method is to obtain a value with a certain number of intervals and then obtain a value with twice the number of intervals. If the resulting values are the same to the required number of decimal places, then this common value can be taken as that required. However this is rather unscientific, and cases can arise where the two values, although the same to the required accuracy, do not give the value of the integral— as doubling the number of intervals yet again might show. An illustrative example due to Clenshaw and Curtis follows.

Example 3.3 Using Simpson's rule with $H = 1, \frac{1}{2}, \frac{1}{4}$ evaluate

$$\tfrac{1}{2} \int_{-1}^{1} (0\cdot 92 \cosh x - \cos x)\, dx$$

and compare, to six decimal places, with the exact value.

The integral is $(0\cdot 92 \sinh 1 - \sin 1) = 0\cdot 239\,714$, correct to six decimal places. Using eq. (3.3) with

(a) $H = 1, m = 2$, (b) $H = \frac{1}{2}, m = 4$, (c) $H = \frac{1}{4}, m = 8$

the Simpson-rule values for the integral are, respectively,

(a) $0\cdot 239\,777$, (b) $0\cdot 239\,778$, (c) $0\cdot 239\,719$.

It is noted that the values with $H = 1$ and $H = \frac{1}{2}$ effectively agree to six figures, although the last two are incorrect.

An assessment of the error involved in any numerical solution of the differential equation is therefore desirable, if not absolutely necessary. We consider

$$\int_{x_0}^{x_1} f(x)\, dx$$

where $x_1 - x_0 = h$.

If we expand $f(x)$ as a Taylor series about $x = x_0$ (assuming that the expansion exists), then

$$f(x) = f_0 + (x - x_0)f_0' + \frac{(x - x_0)^2}{2!} f_0'' + \cdots$$

Hence

$$\int_{x_0}^{x_1} f(x)\, dx = hf_0 + \frac{h^2}{2} f_0' + \frac{h^3}{3\cdot 2!} f_0'' + \cdots$$

But

$$f(x_1) = f_1 = f_0 + hf'_0 + \frac{h^2}{2!}f''_0 + \cdots$$

therefore

$$\frac{h^2}{2}f'_0 = \frac{h}{2}(f_1 - f_0) - \frac{h^3}{2\cdot 2!}f''_0 - \cdots$$

and hence

$$\int_{x_0}^{x_1} f(x)\,dx = \frac{h}{2}(f_0 + f_1) - \frac{h^3}{12}f''_0 + \text{terms of higher degree in } h$$

The first term in the series is just the value of the integral as given by the trapezoidal rule. The error in the integral given by the trapezoidal rule is therefore approximately $-\frac{1}{12}h^3 f''_0$, which is known as the local error. Hence if the integration is extended over n intervals, the approximate global error given by the trapezoidal rule is $-\frac{1}{12}nh^3 f''$, where f'' is a value of the second derivative in the range $x_0 \leq x \leq x_n$. Since $x_n - x_0 = nh$, the global error can be written as $-\frac{1}{12}(x_n - x_0)h^2 f''$. In a similar way, by expansion about $x = x_1$,

$$\int_{x_0}^{x_2} f(x)\,dx = \frac{h}{3}(f_0 + 4f_1 + f_2) - \frac{1}{90}h^5 f_1^{(4)} + \cdots$$

where $x_1 - x_0 = x_2 - x_1 = h$.

Hence the approximate global error given by Simpson's rule over an even number of m intervals is

$$-mh^5 f^{(4)}/180 = -(x_m - x_0)h^4 f^{(4)}/180$$

where $f^{(4)}$ is a value of the fourth derivative in the range $x_0 \leq x \leq x_m$.

Example 3.4 Estimate the errors using the trapezoidal rule for the integral of Example 3.2.

$$f(x) = \frac{1}{1 + x^2}$$

hence

$$f''(x) = \frac{2(3x^2 - 1)}{(1 + x^2)^3}$$

We take f'' to be the largest numerical value of $f''(x)$ in the range $0 \leq x \leq 1$ so that the error is not underestimated; this occurs at $x = 0$ and is -2.

(a) $n = 2, h = \frac{1}{2}$;

hence the error is approximately $-2(0\cdot5)^3\,(-2)/12 = 0\cdot0417$.

(b) $n = 4, h = \frac{1}{4}$;

hence the error is approximately $-4(0\cdot25)^3\,(-2)/12 = 0\cdot0104$.

(c) $n = 8, h = \frac{1}{8}$,

hence the error is approximately $-8(0\cdot125)^3\,(-2)/12 = 0\cdot0026$.

We have stated that to evaluate an integral numerically, it is usual to obtain a value for the integral using a particular integration formula with a particular step size, then to repeat the process with half step size and then perhaps repeat the process with the step size again halved. The question arises whether this sequence of numbers can be used to obtain an even better estimation for the value of the integral. Consider the trapezoidal rule with a step size h. Then the required value of the integral is

$$A = \int_a^b f(\eta)\,d\eta = A_0 - \tfrac{1}{12}(b-a)h^2 f''$$

(if terms involving h^3 are neglected), where A_0 is the approximate value for the integral given by the trapezoidal rule with a step size h. With the step size halved,

$$A = A_1 - (b-a)h^2 f''/48$$

where A_1 is the approximate value for the integral given by the trapezoidal rule with a step size $\frac{1}{2}h$. The assumption is now made that the values of the second derivatives occurring in the error terms are the same for both step sizes. With this assumption we can write

$$A = A_0 + G = A_1 + \tfrac{1}{4}G$$

where G is a constant. Hence

$$G = \tfrac{4}{3}(A_1 - A_0) \qquad \text{and} \qquad A = \tfrac{1}{3}(4A_1 - A_0)$$

The error in this estimate for the value of the integral now depends upon a term involving h^3 rather than h^2. This method of improving the accuracy is known as *Richardson's extrapolation*.

62 Ordinary Differential Equations

A similar extrapolation technique can be applied to other integration formulae.

Example 3.5 Use Richardson's extrapolation on the trapezoidal rule estimates of the values of the integral of Example 3.2 given by step sizes of $\frac{1}{4}$ and $\frac{1}{8}$.

From Example 3.2,

$$A_0 = 0.782\ 794 \qquad A_1 = 0.784\ 747$$

Hence

$$A = \tfrac{1}{3}(4A_1 - A_0) = 0.785\ 398$$

We note that to six decimal places this value agrees with the exact value for the integral.

Problem 3.1 Evaluate $\int_0^1 \sin \pi x\, dx$ using the trapezoidal rule and Simpson's rule with (a) $m = 4$, (b) $m = 8$, (c) $m = 10$. In each case estimate the error and compare with the exact value of the integral to five decimal places.

Problem 3.2 Use Richardson's extrapolation with the trapezoidal rule estimates (a) and (b) of the previous problem to obtain a better numerical value for the integral.

Problem 3.3 Obtain an extrapolation formula for Simpson's rule, and using this formula with appropriate values already obtained in Problem 3.1, find another estimate for the integral of that problem.

3.3 One-step methods

We now consider first-order differential equations of the form

$$\frac{dy}{dx} = f(x, y) \qquad \text{with } y = y_0 \text{ when } x = x_0 \qquad (3.4)$$

We assume that the value of the solution is required at $x = x_n$. If $(x_n - x_0)$ is large, it is necessary to build up the numerical solution by finding the solution at a number of internal points $x_1, x_2, \ldots, x_{n-1}$. We can therefore consider the problem of finding the solution at $x = x_1$

Numerical Solutions of First-Order Differential Equations 63

where the step length $(x_1 - x_0) = h$ can be regarded as small. The process used can then be repeated to find the solution at $x = x_2$ and so on. Care has to be taken that any errors introduced by the numerical evaluation do not build up unacceptably as we proceed from step to step.

We start by considering so called one-step methods. These are methods which proceed from the known solution at x_r to the solution at x_{r+1} by using only the known values at x_r. Having obtained solutions of the differential equation at the values of $x_1, x_2, x_3, \ldots, x_r$, other methods of finding the solution at x_{r+1} use all or some of these values rather than just the solution at x_r. Some such methods are known as predictor–corrector methods, and these are described and discussed in Section 3.4.

3.3.1 Series solution

Solution in series is considered in detail in Chapter 6, but it is very briefly mentioned here for use in the following sections.

If $y(x)$ can be expanded in a Taylor series about $x = x_0$, then

$$y(x) = y_0 + (x - x_0)y_0^{(1)} + \frac{(x - x_0)^2}{2!} y_0^{(2)} + \frac{(x - x_0)^3}{3!} y_0^{(3)} + \ldots$$

where

$$y_0 = y(x_0) \quad \text{and} \quad y^{(n)} = \frac{d^n y}{dx^n} \tag{3.5}$$

But from (3.4), $y^{(1)} = f$, hence on evaluating (3.5) at $x = x_1$, assuming convergence of the series,

$$y(x_1) = y_1 = y_0 + hf_0 + \frac{h^2}{2!} f_0^{(1)} + \frac{h^3}{3!} f_0^{(2)} + \ldots \tag{3.6}$$

Now f_0 is known and $f_0^{(1)}$ can be obtained by differentiating the differential equation (3.4) with respect to x and then substituting $x = x_0$, $y = y_0$. Similarly the other derivatives $f_0^{(n)}$ can be evaluated. Hence the value of y_1 can be obtained to any required accuracy. An example of this is given in Example 6.2. This method, although on the surface appearing quite simple, has the disadvantage that in practice the derivatives $y^{(n)}$ are difficult to obtain.

3.3.2 Runge-Kutta formulae

We illustrate the method by considering firstly the *second order* Runge–Kutta formula.

64 Ordinary Differential Equations

Using the same principle that led to the trapezoidal rule for numerical integration (Section 3.2), we can take an approximate value of the solution at $x = x_1$ to be

$$\bar{y}_1 = y_0 + h[\alpha f(x_0, y_0) + \beta f(x_1, y_1)]$$

where

$$\alpha + \beta = 1.$$

However, the solution y_1 is not known at x_1, so that the value of the derivative of the required solution curve at $x = x_1$, that is $f(x_1, y_1)$, is also not known. We therefore take an approximation for the solution at x_1 to be

$$\bar{y}_1 = y_0 + h[\alpha f(x_0, y_0) + \beta f(x_0 + ah, y_0 + bhf_0)]$$

where $\alpha + \beta = 1$ and where a and b are chosen such that $\bar{y}_1$ is as close an approximation to the exact solution y_1 as is possible.

Using Taylor's theorem for a function of two variables, expansion about $x = x_0$ gives

$$\bar{y}_1 = y_0 + h\left[\alpha f_0 + \beta\left(f_0 + ah\frac{\partial f_0}{\partial x} + bhf_0\frac{\partial f_0}{\partial y} + R\right)\right] \quad (3.7)$$

where $f_0 = f(x_0, y_0)$ and R consists of terms involving h^n, where $n \geq 2$. That is

$$\bar{y}_1 = y_0 + (\alpha + \beta)f_0 h + \left(a\beta\frac{\partial f_0}{\partial x} + b\beta f_0\frac{\partial f_0}{\partial y}\right)h^2 + \cdots \quad (3.8)$$

We now compare (3.8) against the expansion (3.6) for the exact solution y_1.

Since $(\alpha + \beta) = 1$, we note that the constant terms and the terms involving h are the same for each expansion. The terms involving h^2 can be made the same by taking

$$\tfrac{1}{2}f_0' = a\beta\frac{\partial f_0}{\partial x} + b\beta f_0\frac{\partial f_0}{\partial y}$$

But

$$f' = \frac{d}{dx}f[x, y] = \frac{\partial f}{\partial x} + \frac{\partial f}{\partial y}\frac{dy}{dx} = \frac{\partial f}{\partial x} + f\frac{\partial f}{\partial y}$$

on using (3.4). Hence the h^2 terms are equal, provided that

$$\frac{1}{2}\left(\frac{\partial f_0}{\partial x} + f_0\frac{\partial f}{\partial y}\right) = a\beta\frac{\partial f_0}{\partial x} + b\beta f_0\frac{\partial f_0}{\partial y}$$

that is if $\tfrac{1}{2} = a\beta = b\beta$.

Hence $b = a$, $\alpha = (1 - \frac{1}{2}a)$, $\beta = \frac{1}{2}a$. The second-order Runge–Kutta estimate for the solution can therefore be written as

$$\bar{y}_1 = y_0 + (1 - \tfrac{1}{2}a)k_1 + (\tfrac{1}{2}a)k_2 \qquad (3.9)$$

where

$$k_1 = hf(x_0, y_0)$$
$$k_2 = hf(x_0 + ah, y_0 + ak_1)$$

and where the local error is proportional to h^3.

We note that any real value for a can be taken, two values frequently used are either $a = 1$ or $a = \frac{1}{2}$.

The second-order Runge–Kutta estimate for the solution at x_{r+1}, obtained from the known numerical solution at x_r, $(x_{r+1} - x_r = h)$, is therefore, with $a = 1$,

$$\bar{y}_{r+1} = \bar{y}_r + \tfrac{1}{2}(k_1 + k_2) \qquad (3.10)$$

where

$$k_1 = hf(x_r, \bar{y}_r) \quad \text{and} \quad k_2 = hf(x_r + h, \bar{y}_r + k_1)$$

and with $a = \frac{1}{2}$

$$\bar{y}_{r+1} = \bar{y}_r + k_2 \qquad (3.11)$$

where

$$k_1 = hf(x_r, \bar{y}_r) \quad \text{and} \quad k_2 = hf(x_r + \tfrac{1}{2}h, \bar{y}_r + \tfrac{1}{2}k_1)$$

A similar procedure can be used to obtain higher-order approximations, which, as for the second-order case, are not unique. A *third-order* formula is

$$\bar{y}_1 = y_0 + \tfrac{1}{6}(k_1 + 4k_2 + k_3) \qquad (3.12)$$

where

$$k_1 = hf(x_0, y_0)$$
$$k_2 = hf(x_0 + \tfrac{1}{2}h, y_0 + \tfrac{1}{2}k_1)$$
$$k_3 = hf(x_0 + h, y_0 - k_1 + 2k_2)$$

The local error is proportional to h^4.

A *fourth-order* formula is

$$\bar{y}_1 = y_0 + \tfrac{1}{6}(k_1 + 2k_2 + 2k_3 + k_4) \qquad (3.13)$$

where
$$k_1 = hf(x_0, y_0)$$
$$k_2 = hf(x_0 + \tfrac{1}{2}h, y_0 + \tfrac{1}{2}k_1)$$
$$k_3 = hf(x_0 + \tfrac{1}{2}h, y_0 + \tfrac{1}{2}k_2)$$
$$k_4 = hf(x_0 + h, y_0 + k_3)$$

The local error is proportional to h^5.

Example 3.6 Find y when $x = 0.2$, given that $dy/dx = 2xy^2$, with $y = 1$ when $x = 0$, using a third-order Runge–Kutta formula in (a) two equal steps and (b) one step.

$$f(x, y) = 2xy^2$$

(a) *First step* $x_0 = 0$, $y_0 = 1$, $h = 0.1$
Using (3.12), $k_1 = 0.1 f(0, 1) = (0.1)(2)(0)(1)^2 = 0$
$k_2 = 0.1 f(0.05, 1) = (0.1)(2)(0.05)(1)^2 = 0.01$
$k_3 = 0.1 f(0.1, 1.02) = (0.1)\,2(0.1)(1.02)^2 = 0.0208$

and
$$\bar{y}_1 = y_0 + \tfrac{1}{6}(k_1 + 4k_2 + k_3) = 1.0101$$

Second step $x_1 = 0.1$, $\bar{y}_1 = 1.0101$, $h = 0.1$
$k_1 = 0.1 f(0.1, 1.0101) = 0.0204$
$k_2 = 0.1 f(0.15, 1.0203) = 0.0312$
$k_3 = 0.1 f(0.2, 1.0521) = 0.0443$

and
$$\bar{y}_2 = \bar{y}_1 + \tfrac{1}{6}(k_1 + 4k_2 + k_3) = 1.0417$$

(b) $x_0 = 0$, $y_0 = 1$, $h = 0.2$.
$k_1 = 0.2 f(0, 1) = 0$
$k_2 = 0.2 f(0.1, 1) = 0.04$
$k_3 = 0.2 f(0.2, 1.08) = 0.0933$

and
$$\bar{y}_1 = y_0 + \tfrac{1}{6}(k_1 + 4k_2 + k_3) = 1.0422$$

Note that since in this case $h = 0.2$, $\bar{y}_1$ is the estimated solution at $x = 0.2$ and compares with $\bar{y}_2$ using $h = 0.1$. A better measure of the error involved in the Runge–Kutta process can be obtained in a rather similar manner to that used in Richardson's extrapolation for numerical integration. The solution is estimated by taking two equal steps and

Numerical Solutions of First-Order Differential Equations   67

one step. Then if $\bar{y}_2$ is the solution using two steps and $\bar{y}_2^*$ is the solution using a double-length step and assuming that the error in one step of size h of an nth-order Runge–Kutta formula is Kh^{n+1}, where K is a constant,

$$y_2 = \bar{y}_2 + 2Kh^{n+1} \qquad \text{using two steps of size } h$$

and

$$y_2 = \bar{y}_2^* + K(2h)^{n+1} \qquad \text{using one step of size } 2h$$

Hence

$$Kh^{n+1} = (\bar{y}_2 - \bar{y}_2^*)/(2^{n+1} - 2)$$

and this is an estimate of the error of an nth-order Runge–Kutta estimate over one step of length h.

Example 3.7   Find an estimate of the error of the numerical solution at $x = 0.2$ of the differential equation of Example 3.6, using two steps and a third-order Runge–Kutta formula.

From Example 3.6, $\bar{y}_2 = 1.0417$ and $\bar{y}_2^* = 1.0422$. Hence an estimate of the error in $\bar{y}_2$ is, since two steps are involved,

$$2\left(\frac{1.0417 - 1.0422}{2^4 - 2}\right) = -0.000\,07$$

Thus to four decimal places $\bar{y}_2$ is likely to differ by only one unit in the last decimal place from the exact solution.

We note that the analytic solution of the differential equation is $y = 1/(1 - x^2)$. Hence at $x = 0.2$, to four decimal places $y = 1.0417$, agreeing with the above.

Problem 3.4   By expansion in series and comparing with the Taylor series (3.6), find the necessary relationships between the constants, $\alpha, \beta, \gamma, a, b, c, d, e$ such that the formula

$$\bar{y}_1 = y_0 + \alpha k_1 + \beta k_2 + \gamma k_3$$

where

$$k_1 = hf(x_0, y_0)$$
$$k_2 = hf(x_0 + ah, y_0 + bk_1)$$
$$k_3 = hf(x_0 + ch, y_0 + dk_1 + ek_2)$$

is a third-order Runge–Kutta formula.

Using the equations, verify that the following two sets of constants give valid third-order Runge–Kutta formulae;

(a) $\alpha = \frac{1}{6}, \beta = \frac{2}{3}, \gamma = \frac{1}{6}, a = b = \frac{1}{2}, c = 1, d = -1, e = 2$
(b) $\alpha = \frac{1}{4}, \beta = 0, \gamma = \frac{3}{4}, a = b = \frac{1}{3}, c = \frac{2}{3}, d = 0, e = \frac{2}{3}$

Problem 3.5 Using the third-order Runge–Kutta formula (3.12), estimate the solution at $x = 0.2$ using one and two equal steps if

$$dy/dx = x + y, \quad \text{with } y = 1 \text{ when } x = 0$$

Using the two values estimate the error.

3.4 Predictor–corrector methods

The one-step methods considered in the previous sections obtain the solution at the end of a step by using the known solution at the beginning of the step. This process could be repeated to obtain the solution at as many points as required; however, it seems that it might be more efficient to obtain the solution by making use of the solution at many preceding points rather than just the immediately preceding point. Some methods that do this are known as predictor–corrector methods, and are now considered.

3.4.1 Euler's method

The solution at $x = x_1$ is given by

$$y_1 = y_0 + \int_{x_0}^{x_1} f(x, y) \, dx \qquad (3.14)$$

where in the integrand y is a function of x, which is of course not as yet known, except at the point $x = x_0$ when $y = y_0$ and $f(x_0, y_0) = f_0$. The integral can, as has been described in Section 3.2, be approximated by taking various weighted averages of the integrand. Hence a predicted value of the solution at $x = x_1$ can be taken as

$$\bar{y}_1^p = y_0 + hf_0 \qquad (3.15)$$

Comparing this with (3.6), the series expansion for the exact solution y_1, we see that the error is approximately $\frac{1}{2}h^2 f_0^{(1)}$.

Assuming $f(x_1, y_1) = f_1$ is known, then a better approximation for y can be obtained by using the trapezoidal rule for the integration

Numerical Solutions of First-Order Differential Equations   69

involved in (3.14). Hence this corrected value is given by

$$\bar{y}_1^c = y_0 + \tfrac{1}{2}h(f_0 + f_1) \qquad (3.16)$$

The error as given in Section 3.2 is approximately $-\tfrac{1}{12}h^3 f_0^{(2)}$. However, if in the right-hand side of (3.16), f_1 is replaced by $f_1^c = f(x_1, \bar{y}_1^c)$, then the error is not changed as far as the h^3 term is concerned. Hence we take

$$\bar{y}_1^c = y_0 + \tfrac{1}{2}h(f_0 + f_1^c) \qquad (3.17)$$

with error approximately $-\tfrac{1}{12}h^3 f_0^{(2)}$.

The value of f_1^c cannot be obtained directly, and is usually obtained by an iterative procedure which evaluates intermediate values $\bar{y}_1^{c(r)}$, where

$$\bar{y}_1^{c(1)} = y_0 + \tfrac{1}{2} h(f_0 + f_1^p)$$
$$\bar{y}_1^{c(2)} = y_0 + \tfrac{1}{2}h(f_0 + f_1^{c(1)})$$
$$\vdots$$
$$\bar{y}_1^{c(r+1)} = y_0 + \tfrac{1}{2}h(f_0 + f_1^{c(r)})$$
$$\vdots$$

and where

$$f_1^p = f(x_1, \bar{y}_1^p) \qquad f_1^{c(r)} = f(x_1, \bar{y}_1^{c(r)})$$

When the sequence has converged, the solution $\bar{y}_1^c$ has an error $-\tfrac{1}{12}h^3 f_0^{(2)}$. In many cases, however, it is sufficient to obtain $\bar{y}_1^{c(1)}$ as the error is Ah^3, where A is more difficult to obtain but where nevertheless the error is of the same order as the converged value, $\bar{y}_1^c$.

Example 3.8 In one step, find y when $x = 0.6$, given $dy/dx = 1 + y^2$ with $y = 0.546\,30$ when $x = 0.5$.

For the given differential equation, $f(x, y) = 1 + y^2$, so

$$f_0 = f(0.6, 0.546\,30) = 1.298\,44$$

Hence

$\bar{y}_1^p = 0.546\,30 + (0.1)(1.298\,44) = 0.676\,14$

$\bar{y}_1^{c(1)} = 0.546\,30 + 0.05\,[1.298\,44 + (1 + 0.676\,14^2)] = 0.684\,08$

$\bar{y}_1^{c(2)} = 0.546\,30 + 0.05\,[1.298\,44 + (1 + 0.684\,08^2)] = 0.684\,62$

$\bar{y}_1^{c(3)} = 0.546\,30 + 0.05\,[1.298\,44 + (1 + 0.684\,62^2)] = 0.684\,66$

$\bar{y}_1^{c(4)} = 0.546\,30 + 0.05\,[1.298\,44 + (1 + 0.684\,66^2)] = 0.684\,66.$

Thus the estimated solution at $x = 0.6$ is $\bar{y}_1^c = 0.684\ 66$. Although the computation has been performed using five places of decimals to avoid rounding errors, we are not justified in quoting the solution to this number of decimal places, since the error term involves $h^3 = 0.001$. Differentiating f twice we obtain $f^{(2)} = 2f(f + 2y^2)$, so that $f_0^{(2)} = 4.9$. The approximate error in $\bar{y}_1^c$ is therefore $-(0.1)^3\ 4.9/12 = -0.0004$. Using this value for the error, a better estimate for the solution is 0.6843. We note that the effort in iterating to $\bar{y}_1^c$ from the value $\bar{y}_1^{c(1)}$ has not been worth while. The computational effort would have been much more efficiently spent on increasing the accuracy of the solution by taking two steps of half the length.

3.4.2 Milne's method

Euler's method uses values given at $x = x_0$ to obtain the predicted value of y at $x = x_1$, while the corrected value is obtained by using the values at x_0 and the predicted value at x_1. If values of the solution are known for values of x less than x_0, then we know from Section 3.2 that a better approximation for y_1 can be obtained by taking weighted averages of a number of slopes rather than just a particular value. The same also applies for the evaluation of the corrected value.

The orders of the errors for the predicted and corrected values given by Euler's method are different. It is convenient, for a reason to be given shortly, to have the orders of the errors of the two values the same. We now obtain corrected and predicted values which have errors of order h^5.

We assume that values of the solution curve and hence its slope are known at $x_{-3}, x_{-2}, x_{-1}, x_0$, and we wish to find the value at x_1. The length of the interval x_{-3} to x_1 is $4h$, and we estimate the value at x_1 by a straight line passing through the point (x_{-3}, y_{-3}) and with a slope which is some weighted average of those at $x_{-3}, x_{-2}, x_{-1}, x_0$. That is

$$\bar{y}_1^p = y_{-3} + 4h[\alpha f_{-3} + \beta f_{-2} + \gamma f_{-1} + \delta f_0] \qquad (3.18)$$

where

$$\alpha + \beta + \gamma + \delta = 1 \qquad (3.19)$$

We choose the constants $\alpha, \beta, \gamma, \delta$ so that when each of the terms on the right-hand side of eq. (3.18) are expanded as Taylor series about $x = x_0$ then $\bar{y}_1^p$ agrees with the series expansion for the exact solution y_1, eq. (3.6), to as high an order as possible. That is, on substituting

$$y_{-3} = y_0 - 3hf_0 + \tfrac{9}{2}h^2 f_0^{(1)} - \tfrac{9}{2}h^3 f_0^{(2)} + \tfrac{27}{8}h^4 f_0^{(3)} - \tfrac{81}{40}h^5 f_0^{(4)} + \ldots$$
$$f_{-3} = f_0 - 3hf_0^{(1)} + \tfrac{9}{2}h^2 f_0^{(2)} - \tfrac{9}{2}h^3 f_0^{(3)} + \tfrac{27}{8}h^4 f_0^{(4)} - \ldots$$
$$f_{-2} = f_0 - 2hf_0^{(1)} + 2h^2 f_0^{(2)} - \tfrac{4}{3}h^3 f_0^{(3)} + \tfrac{2}{3}h^4 f_0^{(4)} - \ldots$$
$$f_{-1} = f_0 - hf_0^{(1)} + \tfrac{1}{2}h^2 f_0^{(2)} - \tfrac{1}{6}h^3 f_0^{(3)} + \tfrac{1}{24}h^4 f_0^{(4)} - \ldots$$

into (3.18):

$$\begin{aligned}\bar{y}_1^p = y_0 &+ h(-3 + 4\alpha + 4\beta + 4\gamma + 4\delta)f_0 \\ &+ h^2(\tfrac{9}{2} - 12\alpha - 8\beta - 4\gamma)f_0^{(1)} \\ &+ h^3(-\tfrac{9}{2} + 18\alpha + 8\beta + 2\gamma)f_0^{(2)} \\ &+ h^4(\tfrac{27}{8} - 18\alpha - \tfrac{16}{3}\beta - \tfrac{2}{3}\gamma)f_0^{(3)} \\ &+ h^5(-\tfrac{81}{40} + \tfrac{27}{2}\alpha + \tfrac{8}{3}\beta + \tfrac{1}{6}\gamma)f_0^{(4)} + \ldots\end{aligned} \qquad (3.20)$$

Comparing this with (3.6), note that the first two terms agree, on using (3.19), so that α, β, γ can be chosen such that the two expansions agree up to and including the term involving h^4. Solving the appropriate equations,

$$\alpha = 0 \quad \beta = \tfrac{2}{3} \quad \gamma = -\tfrac{1}{3} \quad \delta = \tfrac{2}{3} \qquad (3.21)$$

The required predictor is therefore

$$\bar{y}_1^p = y_{-3} + \tfrac{4}{3}h(2f_0 - f_{-1} + 2f_{-2}) \qquad (3.22)$$

By comparing (3.20) with (3.6), we calculate the error approximately as $28h^5 f_0^{(4)}/90$.

Assuming that $f_1 = f(x_1, y_1)$ is known, another estimate of y_1 can be obtained by using Simpson's rule over the interval x_{-1} to x_1, and is

$$\bar{y}_1^c = y_{-1} + \tfrac{1}{3}h(f_1 + 4f_0 + f_{-1}) \qquad (3.23)$$

The error is known from Section 3.2 to be approximately $-\tfrac{1}{90}h^5 f_0^{(4)}$. Using the same arguments as in the previous section we can take

$$\bar{y}_1^c = y_{-1} + \tfrac{1}{3}h(f_1^c + 4f_0 + f_{-1}) \qquad (3.24)$$

with the same approximate error.

72 Ordinary Differential Equations

It is necessary to iterate to $\bar{y}_1^c$ in a similar way to that described in 3.4.1.

The formulae (3.22) and (3.24) are known as *Milne's formulae*.

The numerical value of the approximate error in $\bar{y}_1^c$ is not easily obtained since the evaluation of $f_0^{(4)}$ can be difficult and time consuming. However, since the solution has been approximated by $\bar{y}_1^p$ and $\bar{y}_1^c$ to the same order of accuracy, the approximate error in $\bar{y}_1^c$ can be estimated as follows.

Neglecting terms involving h^n, $n \geqslant 6$, we have that

$$y_1 = \bar{y}_1^p + \tfrac{28}{90}h^5 f_0^{(4)} \qquad \text{and} \qquad y_1 = \bar{y}_1^c - \tfrac{1}{90}h^5 f_0^{(4)}$$

Hence

$$\bar{y}_1^c - \bar{y}_1^p = \tfrac{29}{90}h^5 f_0^{(4)} \qquad \text{and} \qquad y_1 = \bar{y}_1^c - \tfrac{1}{29}(\bar{y}_1^c - \bar{y}_1^p)$$

The approximate error in $\bar{y}_1^c$ is therefore $-\tfrac{1}{29}(\bar{y}_1^c - \bar{y}_1^p)$, which is obviously very easily evaluated to obtain an improved solution. If the error is too large then the step sizes should be decreased.

Example 3.9 In one step find y when $x = 0 \cdot 6$, given $dy/dx = 1 + y^2$, together with the table of values

x	0	0·1	0·2	0·3	0·4	0·5
y	0	0·100 335	0·202 710	0·309 336	0·422 793	0·546 302

using Milne's predictor–corrector formula.

From the table of values, with $x_0 = 0 \cdot 5$,

$$y_{-3} = y(0 \cdot 2) = 0 \cdot 202\ 710 \qquad y_{-1} = y(0 \cdot 4) = 0 \cdot 422\ 793$$

$$f_{-2} = 1 \cdot 095\ 689 \qquad f_{-1} = 1 \cdot 178\ 754 \qquad f_0 = 1 \cdot 298\ 446$$

Hence from (3.22), $\bar{y}_1^p = 0 \cdot 683\ 979$.

The first corrected estimate is therefore

$$\bar{y}_1^{c(1)} = y_{-1} + \tfrac{1}{3}h[f_1^p + 4f_0 + f_{-1}]$$

where

$$f_1^p = f(x_1, \bar{y}_1^p) = 1 \cdot 467\ 827$$

Hence

$$\bar{y}_1^{c(1)} = 0 \cdot 684\ 139$$

Proceeding in a similar way, we have
$$\bar{y}_1^{c(2)} = 0.684\ 146 \qquad \bar{y}_1^{c(3)} = 0.684\ 146$$
and hence
$$\bar{y}_1^c = 0.684\ 146$$
An estimate of the error in $\bar{y}_1^c$ is given by
$$-\tfrac{1}{29}(\bar{y}_1^c - \bar{y}_1^p) = -0.000\ 006$$
so that the improved solution is $\bar{y}_1 = 0.684\ 140$.

We note again that it is sufficient to estimate the solution y_1 by $\bar{y}_1^{c(1)}$ and that an estimate of its error can be taken to be
$$-\tfrac{1}{29}(\bar{y}_1^{c(1)} - \bar{y}_1^p)$$
As far as computational effort is concerned, predictor–corrector methods are usually less time consuming than Runge–Kutta ones. Therefore, stability considerations apart, it is usual to use Runge–Kutta or series formula to obtain a number of solution values near the given initial condition, then when enough of these are obtained to use predictor–corrector formulae.

Problem 3.6 The values of the solution to the differential equation $dy/dx = x + y$ at a number of values of x are given as follows:

x	0	0·1	0·2	0·3
y	1·000 00	1·110 34	1·242 81	1·399 72

With $h = 0.1$, find an estimate of the solution at $x = 0.4$, using (a) Euler's and (b) Milne's predictor–corrector formulae. Estimate the likely error in both cases.

3.5 Ill-conditioning, stability

We briefly consider how numerical errors introduced by the numerical methods build up as we proceed with the solution step by step.

3.5.1 Ill-conditioning

Consider the solution curves for the differential equation $dy/dx = y - \tfrac{2}{3}$. The solution curves are $y = ae^x + \tfrac{2}{3}$ and are sketched in Fig. 3.2.

Ordinary Differential Equations

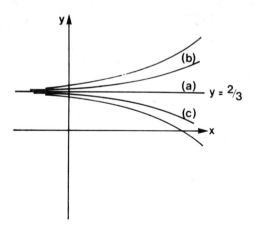

Fig. 3.2

Let us suppose that we require the solution subject to the initial condition $y = \frac{2}{3}$ when $x = 0$. We see that the solution is $y = \frac{2}{3}$, all x, curve (a) in Fig. 3.2. If at any stage in the numerical method used an error were introduced which overestimated the value of y, then the solution would proceed along a solution curve (b), since $dy/dx > 0$. This curve increases exponentially, and we see that the solution obtained would diverge rapidly from the correct values. If the value were underestimated, then the solution would proceed along curve (c).

If, for example, the initial condition were taken correct to two decimal places, that is $y = 0.67$ when $x = 0$, the resulting solution correct to two decimal places would be as given below.

x	0	2	4	6	8	10
y	0·67	0·69	0·85	2·01	10·60	74·09

A differential equation which leads to this behaviour is said to be *ill-conditioned*.

A test for ill-conditioning is obtained by repeating the numerical solution with a slightly different initial condition, and comparing the solutions.

3.5.2 Stability

(a) *Step length*

Let us obtain the solution, at $x = 1$, of the differential equation $dy/dx = -10y$, subject to $y = 1$ when $x = 0$, using Euler's method with n equal intervals.

We have that $h = 1/n$ and $f(x, y) = -10y$; hence $\bar{y}_1^p = y_0 + hf_0 = 1 - (10/n)$. Similarly,

$$\bar{y}_2^p = \bar{y}_1^p + hf_1^p = \left(1 - \frac{10}{n}\right) - \frac{10}{n}\left(1 - \frac{10}{n}\right) = \left(1 - \frac{10}{n}\right)^2$$

and hence, continuing in this way,

$$\bar{y}(1) = \bar{y}_n^p = \left(1 - \frac{10}{n}\right)^n$$

If we take 5 intervals so that $n = 5$, $h = 0.2$, then $\bar{y}_n^p = -1$. This compares with the exact value of $y_5 = e^{-10} = 4.54 \times 10^{-5}$ to three significant figures. When $n = 10$, $\bar{y}_n^p = 0$, when $n = 1000$, $\bar{y}_n^p = 4.32 \times 10^{-5}$, and when $n = 10\,000$, $\bar{y}_n^p = 4.52 \times 10^{-5}$.

We see that for this method and this differential equation, the number of intervals has to be very large, that is, the step length has to be very small, if an acceptable degree of accuracy is to be obtained. Difficulties of this sort arise whenever the solution involves exponentials that decay rapidly to zero, and care has to be taken in the selection of a value for h which gives a satisfactory approximation to the most rapidly decreasing exponential term.

(b) *Parasitic solutions*

We have seen in Section 3.5.1 that a differential equation which has a solution in the form of an increasing exponential leads to an unstable situation if any errors are introduced by the numerical solution. In certain cases, although the differential equation may not be ill-conditioned, the equation which approximates the solution, such as (3.24), may introduce a 'parasitic' solution which may be exponentially increasing and so eventually swamp the required solution. Parasitic solutions are introduced whenever the order of the difference equation for the approximate solution is higher than that of the differential equation from which it was derived.

We illustrate this by considering the solution of the differential equation $dy/dx = -y$ subject to $y = 1$ when $x = 0$, using Milne's corrector.

For this differential equation with $f(x, y) = -y$, that is, $f_n = -y_n$, Milne's corrector, (3.24) is

$$\bar{y}_{n+2} = \bar{y}_n + \tfrac{1}{3}h(-\bar{y}_{n+2} - 4\bar{y}_{n+1} - \bar{y}_n)$$

and hence the difference equation for the approximate solution is

$$(3 + h)\bar{y}_{n+2} + 4h\,\bar{y}_{n+1} - (3 - h)\bar{y}_n = 0$$

The solution of this difference equation can be shown to be

$$\bar{y}_n = A Y_1^n + B Y_2^n$$

where Y_1 and Y_2 are functions of h given by

$$(1 + \tfrac{1}{3}h)Y_1 = -\tfrac{2}{3}h + [1 + \tfrac{1}{3}h^2]^{1/2}$$
$$(1 + \tfrac{1}{3}h)Y_2 = -\tfrac{2}{3}h - [1 + \tfrac{1}{3}h^2]^{1/2}$$

For small values of h, Y_1 and Y_2 are approximately given by

$$Y_1 = 1 - h \qquad Y_2 = -1 - \tfrac{1}{3}h$$

The difference equation is second order, so that two conditions are needed in its solution. The constants A and B are therefore to be determined from the given initial condition, and from the solution at a neighbouring point. The resulting value of A is a number approximately equal to the given initial value y_0, while B is a small number. The term $A Y_1^n$ approximates the required solution while $B Y_2^n$ is the parasitic solution. We note that $|Y_2| > 1$ so that eventually the term $B Y_2^n$ must dominate the numerical solution, however small B is. Also, since Y_2 is negative, $B Y_2^n$ alternates in sign with n, so that the approximate solution $\bar{y}_n$ oscillates about the true solution with increasing amplitudes (Fig. 3.3).

To illustrate, we take $h = 0.2$, $y_0 = 1$ and $y_1 = 0.819$; hence

$$Y_1 = 0.818\,73 \qquad Y_2 = -1.068\,73$$
$$A - 1 = 1.434 \times 10^{-4} \qquad B = 1.434 \times 10^{-4}$$

The values of $A Y_1^n$ and $B Y_2^n$ are compared below against the exact solution, to three significant figures, for a number of values of n.

n	0	5	10	25	50
y (exact)	1·000	0·368	0·135	6.74×10^{-3}	4.54×10^{-5}
$A Y_1^n$	1·000	0·368	0·135	6.74×10^{-3}	4.54×10^{-5}
$|B Y_2^n|$	1.43×10^{-4}	2.00×10^{-4}	2.79×10^{-4}	7.56×10^{-4}	3.98×10^{-3}

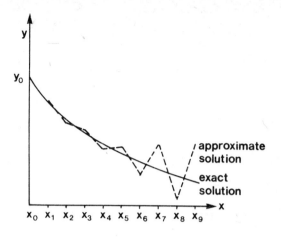

Fig. 3.3

We see from the above set of figures that when the numerical solution has reached $n = 25$, $x = 5$, the parasitic solution is about 10% of the exact value, while when $n = 50$, $x = 10$, the parasitic value completely swamps the exact value.

It can be shown generally that if, from the given initial condition (x_0, y_0), the solution is obtained for *increasing* values of x by Milne's corrector (3.24), then the parasitic solution is unstable if the exact solution is of decaying exponential form, while the parasitic solution disappears if the exact solution is of increasing exponential form. If, from the given condition (x_0, y_0), the solution is obtained for *decreasing* values of x, then the position is completely reversed. Other predictor–corrector systems exist which have parasitic solutions that disappear under all reasonable circumstances. One of these is the *Adams–Bashforth–Moulton method*,

$$\bar{y}_1^p = y_0 + \tfrac{1}{24} h(55 f_0 - 59 f_{-1} + 37 f_{-2} - 9 f_{-3})$$

(3.25)

$$\bar{y}_1^c = y_0 + \tfrac{1}{24} h(9 f_1 + 19 f_0 - 5 f_{-1} + f_{-2})$$

78 Ordinary Differential Equations

Problem 3.7 Obtain an estimate of the error for both the predictor and corrector of the Adams-Bashforth–Moulton formula (3.25). Using the predictor and corrector, obtain an improved solution.

Problem 3.8 Repeat Problem 3.6 using the Adams-Bashforth–Moulton formula.

CHAPTER FOUR

Analytic Solution of Second- and Higher-Order Differential Equations

4.1 Introduction

High-order differential equations are related to systems of differential equations in that any high-order equation can be replaced by a system of lower-order equations, and in particular by a system of first-order equations. We have already illustrated in Sections 1.4.6 and 1.4.7 how second-order equations can be replaced by a pair of first-order equations.

Although we are in this chapter concerned with analytic solution, it should be kept in mind that only the general characteristics of the solution may be required. An example of this was given in Section 1.4.7, where use was made of the phase space. This method of obtaining the general characteristics of the solution is suited to so-called *autonomous* second-order differential equations of the form

$$\frac{d^2 x}{dt^2} = f\left(x, \frac{dx}{dt}\right)$$

where the function f is independent of t. This second-order equation can be replaced by an autonomous first-order system

$$\frac{dx}{dt} = y \qquad \frac{dy}{dt} = f(x, y)$$

80 Ordinary Differential Equations

Since $y(dy/dx) = dy/dt$, the solution curves $y = y(x)$ of the differential equation

$$y\, dy/dx = f(x, y)$$

can then be plotted in the phase space and general characteristics of the solution observed, since for increasing t, as stated in Section 1.4.7, the solutions move from left to right in the upper half plane and from right to left in the lower half plane.

Two examples follow which illustrate how the general characteristics of the solution may be obtained without having to obtain the full analytic or numerical solution.

Example 4.1 Determine the general characteristics of the solution of the differential equation

$$\frac{d^2x}{dt^2} + 4\frac{dx}{dt} + 13x = 0$$

Let $y = dx/dt$, so that the second-order differential equation can be written as a pair of first-order equations

$$\frac{dy}{dt} = -13x - 4y \qquad \frac{dx}{dt} = y$$

Hence

$$\frac{dy}{dx} = -4 - 13\frac{x}{y}$$

From the latter first-order equation in y and x only, the solution curves in the phase space can be obtained either by analytic solution, as described in Section 2.3, or by geometrical construction, as described in Section 1.4.

Since the first-order homogeneous equation leads to a complicated relationship between y and x, we obtain the solution curves by geometrical construction. We note that dy/dx is infinite when $y = 0$, and that it is constant on lines $y = ax$ with the value $(-4 - 13/a)$. The construction of a solution curve is shown in Fig. 4.1.

As shown in Section 1.4.7, the solution proceeds in a clockwise direction as time increases. Hence we see from Fig. 4.1 that x decays to zero and that x oscillates about $x = 0$ with the amplitude of the oscillation tending to zero as time increases.

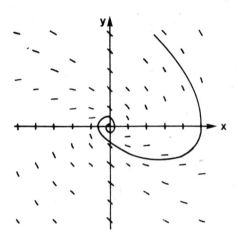

Fig. 4.1

Example 4.2 Determine the general characteristics of the epidemic as modelled by the differential equations derived in Section 1.2.4 when the population is 1000 and initially there are 20 infected people (I), 80 immune people (R), and 900 uninfected people (U). The differential equations (1.15) modelling the epidemic are

$$dR/dt = \alpha I \qquad (4.1)$$
$$dU/dt = -\beta UI \qquad (4.2)$$
$$dI/dt = \beta UI - \alpha I \qquad (4.3)$$

where

$$1000 = U + I + R \qquad (4.4)$$

α and β are given positive constants, and the initial conditions are

$$I(0) = 20 \quad R(0) = 80 \quad U(0) = 900$$

Ordinary Differential Equations

Questions of interest might be:

(a) What is the maximum number of infected people?
(b) At the end of the epidemic how many people have not been infected?

We note that $U, I, R \geqslant 0$, and hence from eqs. (4.1) and (4.2) R increases and U decreases with time.

If T denotes the time at which the epidemic ends then we have that $I(t) = 0$ for $t \geqslant T$, and from eqs. (4.1) and (4.2), $R(t) = $ constant and $U(t) = $ constant for $t \geqslant T$.

From eqs. (4.1) and (4.2)

$$dU/dR = -kU \qquad (4.5)$$

where $k = \beta/\alpha$. The solution of (4.5) is

$$U = U_0 \exp[-k(R - R_0)] \qquad (4.6)$$

where, for the given initial conditions,

$$U_0 = 900 \qquad R_0 = 80$$

The number of uninfected people therefore decreases exponentially as the number of immune people increases.

If suffix E denotes numbers at the end of the epidemic, then since $I_E = 0$

$$1000 = N = R_E + U_E$$

and hence

$$1000 = N = R_E + U_0 \exp[-k(R_E - R_0)]$$

This can be solved numerically to give the following values of R_E and U_E for a few values of k:

k	0	10^{-3}	$1 \cdot 5 \times 10^{-3}$	$2 \cdot 5 \times 10^{-3}$	5×10^{-3}	10^{-2}
R_E	100	209	565	878	991	1000
U_E	900	791	335	122	9	0

The values of U_E answer question (b).

From eqs. (4.1) and (4.3),

$$dI/dR = kU - 1 = k(N - I - R) - 1 \qquad (4.7)$$

Hence the number of infected people increases when $k(N - I - R) > 1$, that is when $kU > 1$, and decreases when $kU < 1$.

Since U is a decreasing function of time, there are two cases to be considered

(a) $kU_0 < 1$

Since the number of infected people always decreases with time, the maximum number of infected people is the initial number I_0 (Fig. 4.2).

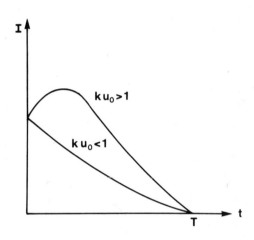

Fig. 4.2

(b) $kU_0 > 1$

The number increases and then decreases so that the maximum is given when $dI/dR = 0$, that is when $kU = 1$ (Fig. 4.2). But

$$kI = kN - kU - kR$$
$$= kN - kU - kR_0 + \ln(kU_0/kU)$$

from eq. (4.6). Hence, with $kU = 1$,

$$kI_{max} = kN - 1 - kR_0 + \ln kU_0$$

84 Ordinary Differential Equations

Values of I_{max} for some k are given below for the given initial conditions:

k	0	10^{-3}	$1 \cdot 5 \times 10^{-3}$	$2 \cdot 5 \times 10^{-3}$	5×10^{-3}	10^{-2}
I_{max}	20	20	53	196	419	600

These figures give the answer to question (a).

Problem 4.1 Sketch in the phase space the solution curves of the differential equations

(a) $d^2 x/dt^2 + n^2 x = 0$ (b) $d^2 x/dt^2 - n^2 x = 0$

and hence describe the motions given by these equations.

Problem 4.2 Sketch in the phase space the solution curves of the differential equation

$$d^2 x/dt^2 + k \, dx/dt + k^2 x = 0$$

for (a) $k > 0$, and (b) $k < 0$.
Hence describe the motion for both cases.

4.2 Homogeneous second-order linear equations with constant coefficients

We consider differential equations of the form

$$a \frac{d^2 x}{dt^2} + b \frac{dx}{dt} + cx = 0 \qquad (4.8)$$

where a, b, c are constants and $a \neq 0$.
Such equations are called *homogeneous* second-order differential equations with constant coefficients.

Theorem If $x_1(t)$ and $x_2(t)$ are solutions of the differential equation (4.8) then

$$x_3(t) = A x_1(t) + B x_2(t)$$

where A and B are any constants, is also a solution.

Proof

$$x_3' = A x_1' + B x_2'$$

and
$$x_3'' = Ax_1'' + Bx_2''$$
Hence
$$ax_3'' + bx_3' + cx_3 = A[ax_1'' + bx_1' + cx_1] + B[ax_2'' + bx_2' + cx_2] = 0$$
since $x_1(t)$ and $x_2(t)$ are solutions of the differential equation (4.8). Hence $x_3(t)$ is also a solution.

The theorem obviously also holds true for any nth-order homogeneous linear differential equation.

Since the solution of the first-order equation $b\,dx/dt + cx = 0$ is $x = \exp(nt)$, where $n = -c/b$ is a constant, we investigate whether there is a solution of (4.8) of the form

$$x = e^{mt} \tag{4.9}$$

where m is a constant to be determined.

With $x = e^{mt}$, $dx/dt = me^{mt}$ and $d^2x/dt^2 = m^2 e^{mt}$. On substituting these into the left-hand side of eq. (4.8), we obtain

$$(am^2 + bm + c)e^{mt}$$

Hence e^{mt} is a solution provided that

$$(am^2 + bm + c)e^{mt} = 0$$

or, since $e^{mt} \neq 0$,

$$am^2 + bm + c = 0 \tag{4.10}$$

This equation for m is known as the *characteristic equation*. Since the characteristic equation is a quadratic, there are two roots which fall into one of three categories:

(a) $b^2 > 4ac$, the roots are unequal and real,
(b) $b^2 < 4ac$, the roots are unequal and complex,
(c) $b^2 = 4ac$, the roots are equal (and real).

We consider each of these cases in turn.

(a) Roots unequal and real, m_1 and m_2.
Then $e^{m_1 t}$ and $e^{m_2 t}$ are solutions of eq. (4.8), and it follows from the theorem that

$$x(t) = Ae^{m_1 t} + Be^{m_2 t} \tag{4.11}$$

86 Ordinary Differential Equations

is also a solution. Since this solution contains two arbitrary constants, it is the required general solution.

(b) Roots complex, $\gamma \pm i\delta$.

By analogy with case (a), the general solution is

$$x(t) = Ae^{(\gamma + i\delta)t} + Be^{(\gamma - i\delta)t}$$

This expression can, however, be rewritten as follows to avoid the use of the imaginary number i.

$$\begin{aligned}x(t) &= e^{\gamma t}[Ae^{i\delta t} + Be^{-i\delta t}] \\ &= e^{\gamma t}[A(\cos \delta t + i \sin \delta t) + B(\cos \delta t - i \sin \delta t)] \\ &= e^{\gamma t}[(A + B)\cos \delta t + (Ai - Bi)\sin \delta t] \\ &= e^{\gamma t}[C \cos \delta t + D \sin \delta t]\end{aligned}$$

where C and D are again constants. Hence, in this case, the general solution is

$$x(t) = e^{\gamma t}[C \cos \delta t + D \sin \delta t] \qquad (4.12)$$

(c) Equal roots, m_1.

In this case we need another independent solution so that the general solution can be obtained. We consider $x(t) = te^{m_1 t}$ and investigate whether it is a solution.

With $x(t) = te^{m_1 t}$

$$dx/dt = m_1 t e^{m_1 t} + e^{m_1 t}$$

and

$$d^2x/dt^2 = m_1^2 t e^{m_1 t} + 2m_1 e^{m_1 t}$$

On substituting these into the left-hand side of (4.8), we obtain

$$te^{m_1 t}(am_1^2 + bm_1 + c) + e^{m_1 t}(2am_1 + b) \qquad (4.13)$$

But $am_1^2 + bm_1 + c = 0$, since m_1 is a root of the characteristic equation. Since also $b^2 = 4ac$, $m_1 = -b/2a$, so that $2am_1 + b = 0$. It follows that expression (4.13) is zero and therefore $te^{m_1 t}$ is a solution to (4.8). The general solution is therefore

$$x(t) = (A + Bt)e^{m_1 t} \qquad (4.14)$$

Analytic Solution of Second- and Higher-Order Differential Equations

Example 4.3 Obtain the general solutions of the following differential equations:

(a) $d^2x/dt^2 - x = 0$
(b) $d^2x/dt^2 - dx/dt - 6x = 0$
(c) $d^2x/dt^2 + x = 0$
(d) $d^2x/dt^2 + 4\,dx/dt + 13x = 0$
(e) $d^2x/dt^2 + 4\,dx/dt + 4x = 0$.

(a) The characteristic equation is

$$m^2 - 1 = 0$$

with roots $m = \pm 1$. Hence the general solution is

$$x = Ae^t + Be^{-t}$$

(b) The characteristic equation is

$$m^2 - m - 6 = 0$$

with roots $m = 3$ and -2. Hence the general solution is

$$x = Ae^{3t} + Be^{-2t}$$

(c) The characteristic equation is

$$m^2 + 1 = 0$$

with roots $m = \pm i$. Hence the general solution is

$$x = e^{0 \cdot t}(A \sin t + B \cos t) = A \sin t + B \cos t$$

This is the same solution as was obtained in Section 1.4.7 by geometrical construction.

(d) The characteristic equation is

$$m^2 + 4m + 13 = 0$$

with roots

$$m = \frac{-4 \pm \sqrt{(16 - 4.13)}}{2} = -2 \pm 3i$$

The general solution is therefore

$$x = e^{-2t}(A \cos 3t + B \sin 3t)$$

The solution can also be written as

$$x = ae^{-2t} \sin(3t + \alpha)$$

and from this we see that x decreases to zero as time increases and that it oscillates about $x = 0$ with decreasing amplitude.

The analytic solution confirms the general characteristics of the solution of this differential equation, which were obtained in Example 4.1.

(e) The characteristic solution is

$$m^2 + 4m + 4 = 0$$

with a double root $m = -2$. The general solution is therefore

$$x = (A + Bt)e^{-2t}$$

Problem 4.3 Obtain (a) the general solutions, and (b) the solutions subject to the given conditions, of the following differential equations:

(i) $d^2x/dt^2 - 3\, dx/dt - 4x = 0$; $x(0) = 1$, $x \to 0$ as $t \to \infty$
(ii) $2\, d^2x/dt^2 + 2\, dx/dt + x = 0$; $x(0) = 0$, $x'(0) = 1$
(iii) $4\, d^2x/dt^2 - 12\, dx/dt + 9x = 0$; $x(0) = x(1) = 1$

4.3 General second-order linear equations with constant coefficients

We consider differential equations of the form

$$a\frac{d^2x}{dt^2} + b\frac{dx}{dt} + cx = f(t) \tag{4.15}$$

where a, b, c are constants, $a \neq 0$.
The general solution $x_c(t)$ of the associated homogeneous differential equation

$$a\frac{d^2x}{dt^2} + b\frac{dx}{dt} + cx = 0$$

is called the *complementary function* for (4.15). Any solution $x_p(t)$ to the differential equation (4.15) is called a *particular integral* for (4.15).

Theorem The general solution of the differential equation (4.15) is the sum of the complementary function and a particular integral.

Analytic Solution of Second- and Higher-Order Differential Equations

Proof. By their definitions,
$$ax_c'' + bx_c' + cx_c = 0$$
and
$$ax_p'' + bx_p' + cx_p = f(t)$$

If these equations are added, the result is
$$a(x_c + x_p)'' + b(x_c + x_p)' + c(x_c + x_p) = f(t)$$

and hence $(x_c + x_p)$ is a solution to the differential equation (4.15). But x_c contains two arbitrary constants and therefore so does $(x_c + x_p)$, which is therefore the required general solution.

The theorem also applies to any nth-order linear differential equation with constant coefficients.

Example 4.4 Find the general solution of
$$d^2 x/dt^2 - x = 1$$

The complementary function has been evaluated in Example 4.3(a), and is
$$x_c(t) = Ae^t + Be^{-t}$$

It is easily seen that $x = -1$ is a solution of the differential equation of this example. Hence $x_p(t) = -1$, and the required general solution is
$$x(t) = Ae^t + Be^{-t} - 1$$

Section 4.2 has dealt with the evaluation of the complementary function, and so it remains to deal with the particular integral. We consider the method of *undetermined coefficients*. The method consists of 'guessing' a particular integral involving unknown constants. On substitution into the differential equation and equating like terms on both sides of the equation a system of equations involving the constants can be obtained. If the guessed form for the particular integral is correct, then the equations can be solved, and hence a particular integral obtained. Although the form of the particular integral has to be guessed, certain guidelines can be laid down. We note that if a polynomial in t is differentiated, then the result is also a polynomial. Hence if the right-hand side of eq. (4.15), $f(t)$, is a polynomial in t, then it is likely that a polynomial of the same order can be found as a particular integral. Similarly if an exponential is differentiated, then the result is also an exponential of the same type, so that it is likely that an exponential of the same type can be found as the particular integral; and if the right-hand side involves $\sin nt$ and/or $\cos nt$, then a particular integral can also be found using $\sin nt$ and $\cos nt$.

It should be stressed at this point that no worries need be entertained about obtaining an erroneous particular integral by this method, since if the wrong form for the particular integral is guessed, the equations for the constants either cannot be formed or cannot be solved.

If $f(t)$ is the sum of two or more different types of functions $f_1(t), f_2(t), \ldots$, then since the differential equation is linear, the particular integral is the sum of the particular integrals of the differential equation with right-hand sides $f_1(t), f_2(t), \ldots$.

If $f(t)$ is the product of two functions $f_1(t)$ and $f_2(t)$, then we guess the form for its particular integral to be the product of the form of the particular integral for $f_1(t)$ and the form of the particular integral for $f_2(t)$.

Before illustrating the method by a number of examples, we summarise the form to be taken for the particular integral by various types of function, $f(t)$.

(a) $f(t)$, polynomial of order n.
    Take
$$x_p(t) = a_n t^n + a_{n-1} t^{n-1} + \cdots + a_1 t + a_0$$

(b) $f(t) = \alpha \sin nt + \beta \cos nt$
    take
$$x_p(t) = a \sin nt + b \cos nt$$

(c) $f(t) = \alpha e^{nt}$
    take
$$x_p(t) = a e^{nt}$$

Example 4.5 Obtain the particular integral of the differential

Analytic Solution of Second- and Higher-Order Differential Equations   91

equation

$$d^2x/dt^2 - dx/dt - 6x = f(t)$$

where $f(t)$ is given by

(a) $t + 1$ (b) $\sin 2t$ (c) $t \sin t$ (d) e^t (e) e^{-2t}

(a) Try $x_p(t) = at + b$. Hence

$$dx_p/dt = a \qquad d^2x/dt^2 = 0$$

so that on substitution into the differential equation,

$$-a - 6(at + b) = t + 1$$

For this to be satisfied for all values of t, we require

$$-6a = 1 \qquad -a - 6b = 1$$

These two equations can be solved for the constants a, b, with the result $a = -\frac{1}{6}$, $b = -\frac{5}{36}$. The required particular integral is therefore

$$x_p(t) = -\tfrac{1}{36}(6t + 5)$$

(b) Try $x_p(t) = a \sin 2t + b \cos 2t$. Hence

$$dx_p/dt = 2a \cos 2t - 2b \sin 2t$$

and

$$d^2x_p/dt^2 = -4a \sin 2t - 4b \cos 2t$$

So that we require

$$(-4a \sin 2t - 4b \cos 2t) - (2a \cos 2t - 2b \sin 2t) - 6(a \sin 2t + b \cos 2t) = \sin 2t$$

For this to be satisfied for all t, we require the coefficient of $\sin 2t$ to be the same on both sides of the equation and similarly for $\cos 2t$. Hence

$$-10a + 2b = 1 \qquad -2a - 10b = 0$$

and so $a = -\frac{5}{52}$, $b = \frac{1}{52}$. The required particular integral is therefore

$$x_p(t) = \tfrac{1}{52}(-5 \sin 2t + \cos 2t)$$

(c) Try

$$x_p(t) = (at + b)(c \sin t + d \cos t)$$
$$= At \sin t + Bt \cos t + C \sin t + D \cos t$$

Hence

$$dx_p/dt = At\cos t - Bt\sin t + (A - D)\sin t + (B + C)\cos t$$

and

$$d^2 x_p/dt^2 = -At\sin t - Bt\cos t - (2B + C)\sin t + (2A - D)\cos t$$

Substituting into the differential equation and collecting like terms, we have

$$t\sin t\,(-7A + B) + t\cos t\,(-A - 7B)$$
$$+ \sin t\,(-A - 2B - 7C + D) + \cos t(2A - B - C - 7D) = t\sin t$$

Hence on equating coefficients of like terms

$$-7A + B = 1$$
$$-A - 7B = 0$$
$$-A - 2B - 7C + D = 0$$
$$2A - B - C - 7D = 0$$

These four equations can be solved to give

$$A = -\tfrac{7}{50} \qquad B = \tfrac{1}{50} \qquad C = \tfrac{1}{125} \qquad D = -\tfrac{11}{250}$$

The required particular integral is therefore

$$x_p(t) = (\tfrac{1}{250})(-35t\sin t + 5t\cos t + 2\sin t - 11\cos t)$$

(d) Try $x_p(t) = ae^t$. Hence

$$dx_p/dt = d^2 x_p/dt^2 = ae^t$$

On substituting into the differential equation, we have

$$-6ae^t = e^t$$

so that we require $a = -\tfrac{1}{6}$. The required particular integral is therefore

$$x_p(t) = -\tfrac{1}{6}e^t$$

(e) Try $x_p(t) = ae^{-2t}$. Hence

$$dx_p/dt = -2ae^{-2t} \qquad d^2 x_p/dt^2 = 4ae^{-2t}$$

On substituting into the differential equation, we have

$$4ae^{-2t} + 2ae^{-2t} - 6ae^{-2t} = e^{-2t}$$

Analytic Solution of Second- and Higher-Order Differential Equations 93

We see that the left-hand side of this equation is identically zero, so that it is not possible to set up an equation for a, and hence it is not possible to obtain a particular integral of the form ae^{-2t}.

The reason for this is that ae^{-2t} is a solution of the associated homogeneous differential equation

$$d^2x/dt^2 - dx/dt - 6x = 0$$

that is, it is one of the independent solutions appearing in the complementary function

$$x_c(t) = Ae^{-2t} + Be^{3t}$$

and so on substituting ae^{-2t} into the left-hand side of the equation, the result must be zero.

The same type of failure of the method occurs whenever any part of $f(t)$ is one of the independent solutions appearing in the complementary function.

In special cases such as these, we take as the guessed particular integral the form of the particular integral that would normally be taken, multiplied by t.

Example 4.5(e) (continued) Since e^{-2t} is a part of the complementary function, we take

$$x_p(t) = ate^{-2t}$$

Hence

$$dx_p/dt = -2ate^{-2t} + ae^{-2t} \quad \text{and} \quad d^2x_p/dt^2 = 4ate^{-2t} - 4ae^{-2t}$$

On substituting into the differential equation, we have

$$te^{-2t}(4a + 2a - 6a) + e^{-2t}(-4a - a) = e^{-2t}$$

We note that the coefficient of te^{-2t} is zero, so that the equation can be satisfied for all values of t if

$$-5a = 1$$

Hence the required particular integral is

$$x_p(t) = -\tfrac{1}{5}te^{-2t}$$

94 Ordinary Differential Equations

The rule for the special case can be generalised for nth order linear equations (see Section 4.4) and summarised:

If the terms on the right-hand side are already contained in the complementary function, corresponding to an r-fold root of the characteristic equation, the expressions given on p. 90 for the particular integral must be multiplied by t^r.

In the solution of differential equations (4.15) it is therefore desirable to evaluate the complementary function before the particular integral.

Example 4.6 Find the particular integral for each of the following differential equations:

(a) $d^2x/dt^2 + x = \sin t$
(b) $d^2x/dt^2 + 4\,dx/dt + 13x = e^{-2t} \sin 3t$
(c) $d^2x/dt^2 + 4\,dx/dt + 4x = e^{-2t}$

(a) The complementary function evaluated in Example 4.3(c) is

$$x_c(t) = A \cos t + B \sin t$$

hence try

$$x_p(t) = t(a \cos t + b \sin t)$$

On substituting into the equation, the values $a = -\tfrac{1}{2}$, $b = 0$ result.

(b) The complementary function evaluated in Example 4.3(d) is

$$x_c(t) = e^{-2t}(A \cos 3t + B \sin 3t)$$

hence try

$$x_p(t) = te^{-2t}(a \cos 3t + b \sin 3t)$$

On substituting into the equation, the values $a = -\tfrac{1}{6}$, $b = 0$ result.

(c) The complementary function evaluated in Example 4.3(e) is

$$x_c(t) = (A + Bt)e^{-2t}$$

Hence try

$$x_p(t) = at^2 e^{-2t}$$

On substituting into the equation, the value $a = \tfrac{1}{2}$ results.

Example 4.7 Find the current and charge in the RLC-series circuit, Fig. 4.3, consisting of a resistance R of 20 ohms, an induction L of 10 H and a capacitor C of 0·05 F if the electromotive force $E(t)$ provided by the battery is $20(1 - \cos t)$ V and if initially at time $t = 0$ the current and charge are zero.

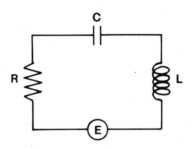

Fig. 4.3

Current I is the rate of change of charge Q so that $I = dQ/dt$. The voltage drop across the resistor is $RI = 20 \, dQ/dt$; that across the induction is $L \, dI/dt = 10 \, d^2Q/dt^2$; that across the capacitor is $Q/C = 20 \, Q$; that across the battery is $-E(t) = -20(1 - \cos t)$. Kirchhoff's second law states that the total voltage drop around the circuit is zero, hence the resulting differential equation is

$$Q'' + 2Q' + 2Q = 2(1 - \cos t)$$

with $Q(0) = 0$ and $I(0) = Q'(0) = 0$. The characteristic equation is

$$m^2 + 2m + 2 = 0$$

with roots $m = -1 \pm i$. Hence the complementary function is

$$Q_c(t) = e^{-t}(A \cos t + B \sin t)$$

For the particular integral, try

$$Q_p(t) = a + b \cos t + c \sin t$$

and on substituting into the differential equation the values $a = 1$, $b = -\frac{2}{5}$, $c = -\frac{4}{5}$ are obtained. The general solution is therefore

$$Q(t) = e^{-t}(A \cos t + B \sin t) + 1 - \tfrac{2}{5}(\cos t + 2 \sin t)$$

The constants A and B are determined from the given initial conditions

$$Q(0) = Q'(0) = 0$$

Differentiating with respect to t,

$$Q'(t) = e^{-t}[(-A + B) \cos t - (A + B) \sin t] + \tfrac{2}{5}(\sin t - 2 \cos t)$$

Hence

$$0 = A + 1 - \tfrac{2}{5}, \quad \text{and} \quad 0 = -A + B - \tfrac{4}{5}$$

so that $A = -\tfrac{3}{5}$ and $B = \tfrac{1}{5}$. The charge at any time t is therefore

$$5Q(t) = e^{-t}(\sin t - 3 \cos t) + 5 - 2(\cos t + 2 \sin t)$$

and the current is

$$5I(t) = 2e^{-t}(2 \cos t + \sin t) + 2(\sin t - 2 \cos t)$$

We note that when the time becomes very large the current is approximately given by

$$I(t) = \tfrac{2}{5}(\sin t - 2 \cos t) = -\tfrac{2}{5}\sqrt{5} \cos(t + \alpha),$$

where $\alpha = \tan^{-1}\tfrac{1}{2}$. This current is called the *steady state current*, and we note that it has the same frequency as that of the electromotive force.

Example 4.8 The output, $x(t)$, of a vibrating system is related to the input, $A \sin \omega t$, by the differential equation

$$\frac{d^2 x}{dt^2} + 2c \frac{dx}{dt} + x = A \sin \omega t$$

where the damping coefficient $c > 0$.

Determine the restraint on c which ensures that the ratio of the amplitude of the steady-state output to the amplitude of the input is less than 4, for all values of the frequency ω.

The characteristic equation is

$$m^2 + 2cm + 1 = 0$$

with roots $m = -c \pm \sqrt{(c^2 - 1)}$. Three cases therefore arise:

(a) $0 < c < 1$.
$$x_c(t) = e^{-ct}[\alpha \sin\sqrt{(1 - c^2)}t + \beta \cos\sqrt{(1 - c^2)}t]$$

(b) $c > 1$
$$x_c(t) = \alpha e^{[-c + \sqrt{(c^2 - 1)}]t} + \beta e^{[-c - \sqrt{(c^2 - 1)}]t}$$

(c) $c = 1$
$$x_c(t) = e^{-ct}(\alpha + \beta t)$$

The particular integral is
$$x_p(t) = a \sin \omega t + b \cos \omega t$$
where
$$a = \frac{(1 - \omega^2)A}{(1 - \omega^2)^2 + 4c^2\omega^2} \qquad b = -\frac{2c\omega A}{(1 - \omega^2)^2 + 4c^2\omega^2}$$

The complementary function decays to zero as the time becomes large for all $c > 0$. The steady-state solution is therefore
$$x(t) = \frac{(1 - \omega^2)A \sin \omega t}{(1 - \omega^2)^2 + 4c^2\omega^2} - \frac{2c\omega A \cos \omega t}{(1 - \omega^2)^2 + 4c^2\omega^2} = B \sin(\omega t - \phi)$$
where
$$B = \frac{A}{\sqrt{[(1 - \omega^2)^2 + 4c^2\omega^2]}} \qquad \text{and} \qquad \phi = \tan^{-1}\frac{2c\omega}{1 - \omega^2}$$

The amplification ratio, A.R. (the ratio of the amplitudes of the steady state output and the input), is therefore
$$\text{A.R.} = \frac{1}{\sqrt{[(1 - \omega^2)^2 + 4c^2\omega^2]}}$$

A sketch of A.R. as a function of the frequency ω is given in Fig. 4.4 for a few values of the damping coefficient c.

It is fairly easily calculated that if $c < 1/\sqrt{2}$ the amplitude ratio has a maximum value of $1/2c\sqrt{(1 - c^2)}$ at a frequency of $\sqrt{(1 - 2c^2)}$, while if $c \geq 1/\sqrt{2}$ the amplitude ratio has a maximum value of unity at zero frequency.

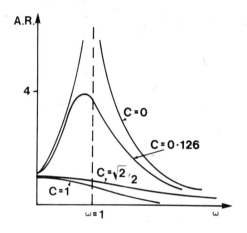

Fig. 4.4

Hence if the amplitude ratio is to be less than four for all frequencies we require

$$4 > \frac{1}{2c\sqrt{(1-c^2)}}$$

that is, $c > 0{\cdot}126$.

This type of problem has important practical applications, since, for instance, cars, aeroplanes, ships and bridges can be regarded as vibrating systems which need to be designed so that they do not build up too large a displacement under any forced input.

Problem 4.4 Find a particular integral for the differential equation

$$d^2x/dt^2 - dx/dt - 2x = f(t)$$

where $f(t)$ is given by

(a) t (b) e^t (c) e^{-t} (d) $\sin t$ (e) te^{-t}

Problem 4.5 Find a particular integral for the differential equation
$$d^2x/dt^2 + 2dx/dt + 2x = e^t \sin t$$

4.4 Higher-order linear equations with constant coefficients

The methods developed in Sections 4.2 and 4.3 for second-order differential equations with constant coefficients can be extended in a straightforward manner to deal with higher-order equations.

Example 4.9 Find the general solution of the differential equation
$$d^3y/dx^3 + 3d^2y/dx^2 + 4dy/dx + 2y = 2x + 1$$

The characteristic equation is
$$m^3 + 3m^2 + 4m + 2 = 0$$
that is,
$$(m + 1)(m^2 + 2m + 2) = 0$$

Hence the roots of the characteristic equation are $-1, -1 \pm i$. The complementary function is therefore
$$y_c(x) = Ae^{-x} + e^{-x}(B \cos x + C \sin x)$$

Since the right-hand side of the equation is a polynomial, we try a particular integral of form
$$y_p(x) = ax + b$$

Differentiating and substituting into the differential equation
$$4a + 2(ax + b) = 2x + 1$$

Hence we require
$$2a = 2 \quad \text{and} \quad 4a + 2b = 1$$
that is, $a = 1$, $b = -\frac{3}{2}$.
The general solution is therefore
$$y(x) = e^{-x}(A + B \cos x + C \sin x) + x - \frac{3}{2}$$

Example 4.10 Find the general solution of the differential equation

$$d^4x/dt^4 - 2\,d^3x/dt^3 + 2\,dx/dt - x = te^t$$

The characteristic equation is

$$m^4 - 2m^3 + 2m - 1 = 0$$

or

$$(m+1)(m-1)^3 = 0$$

That is, there is a single root at -1 and a triple root at 1. Hence the complementary function is

$$x_c(t) = Ae^{-t} + (B + Ct + Dt^2)e^t$$

Since the right-hand side of the differential equation is a part of the complementary function associated with the triple root of the characteristic equation, we take the particular integral to be

$$x_p(t) = t^3(at + b)e^t$$

On substituting into the differential equation, we find the values $a = \frac{1}{48}$ and $b = -\frac{1}{24}$.

The general solution is therefore

$$x(t) = Ae^{-t} + (B + Ct + D^2)e^t + \tfrac{1}{48} t^3(t - 2)e^t$$

Another method of solving nth order linear equations is described in Chapter 7.

Problem 4.6 Find the general solutions to the following differential equations:

(a) $d^3x/dt^3 - 2d^2x/dt^2 = 4$
(b) $d^4x/dt^4 - x = \sin 2t$
(c) $d^3x/dt^3 + 3\,d^2x/dt^2 + 4\,dx/dt + 2x = 0$

4.5 Systems of linear equations with constant coefficients

The solution of a system of simultaneous differential equations with constant coefficients can be obtained by an extension of the methods used for a single equation in Sections 4.2–4.4. That is, the general

Analytic Solution of Second- and Higher-Order Differential Equations

solution to the associated set of homogeneous equations is added to a particular solution.

The approach is illustrated by a number of examples, dealing first with systems of homogeneous equations.

Example 4.11 Determine the general solution to the pair of differential equations

$$dx/dt + 2\, dy/dt - 3x + 4y = 0$$
$$2dx/dt + dy/dt + 2x - y = 0$$

We try for a solution of the form

$$x = ae^{\lambda t} \qquad y = be^{\lambda t}$$

Differentiating and substituting into the differential equation, we obtain the following pair of equations.

$$a(\lambda - 3) + b(2\lambda + 4) = 0 \tag{4.16}$$
$$a(2\lambda + 2) + b(\lambda - 1) = 0 \tag{4.17}$$

From the theory of algebraic equations a non-trivial solution for a and b can be obtained provided that

$$\begin{vmatrix} (\lambda - 3) & (2\lambda + 4) \\ (2\lambda + 2) & (\lambda - 1) \end{vmatrix} = 0$$

where the second-order *determinant* is defined by

$$\begin{vmatrix} c & d \\ e & f \end{vmatrix} = cf - de$$

Hence

$$(\lambda - 3)(\lambda - 1) - (2\lambda + 2)(2\lambda + 4) = 0$$

(For those readers not familiar with the theory of determinants, the same result can be obtained by finding a in terms of b from the first equation and then substituting into the second.) On simplifying, the characteristic equation becomes

$$3\lambda^2 + 16\lambda + 5 = 0$$

which has roots $\lambda = -\frac{1}{3}$ and $\lambda = -5$.

With $\lambda = -\frac{1}{3}$, we have from either of eqs. (4.16) and (4.17) that

$b = a$ and hence a solution pair is

$$x = ae^{-t/3} \qquad y = ae^{-t/3}$$

With $\lambda = -5$ we have from either of eqs. (4.16) and (4.17) that $b = -\tfrac{4}{3}a$, and hence a solution pair is

$$x = ae^{-5t} \qquad y = -\tfrac{4}{3}ae^{-5t}$$

Since the set of equations is linear and homogeneous, we have that if $x_1(t), y_1(t)$ and $x_2(t), y_2(t)$ are two pairs of solutions, then

$$Ax_1(t) + Bx_2(t) \qquad Ay_1(t) + By_2(t)$$

where A and B are any constants, is also a pair of solutions (by an extension of the theorem in Section 4.2).

The required general solution is therefore

$$x(t) = Ae^{-t/3} + Be^{-5t} \qquad y(t) = Ae^{-t/3} - \tfrac{4}{3}Be^{-5t}$$

since it can be shown that the number of independent arbitrary constants appearing in the solution is equal to the order of the characteristic equation.

Example 4.12 Determine the general solution to the set of simultaneous differential equations

$$dx/dt + dy/dt + y = 0$$
$$dx/dt - dz/dt + 2x + z = 0$$
$$dy/dt + dz/dt + y + 2z = 0$$

With $x = ae^{\lambda t}, y = be^{\lambda t}, z = ce^{\lambda t}$, the following equations are obtained:

$$a\lambda \qquad + b(\lambda + 1) \qquad\qquad\qquad = 0 \qquad (4.18)$$
$$a(\lambda + 2) \qquad\qquad + c(-\lambda + 1) = 0 \qquad (4.19)$$
$$\qquad\qquad b(\lambda + 1) + c(\lambda + 2) = 0 \qquad (4.20)$$

A non-trivial solution is obtained if

$$\begin{vmatrix} \lambda & \lambda + 1 & 0 \\ \lambda + 2 & 0 & -\lambda + 1 \\ 0 & \lambda + 1 & \lambda + 2 \end{vmatrix} = 0$$

Analytic Solution of Second- and Higher-Order Differential Equations 103

where the third-order determinant is defined by

$$\begin{vmatrix} c & d & e \\ f & g & h \\ i & j & k \end{vmatrix} = c \begin{vmatrix} g & h \\ j & k \end{vmatrix} - d \begin{vmatrix} f & h \\ i & k \end{vmatrix} + e \begin{vmatrix} f & g \\ i & j \end{vmatrix}$$

$$= cgk + dhi + efj - cjh - dfk - eig$$

Hence the characteristic equation is

$$-(\lambda + 1)(-\lambda + 1)\lambda - (\lambda + 2)(\lambda + 2)(\lambda + 1) = 0$$

that is

$$(\lambda + 1)(5\lambda + 4) = 0$$

with roots $\lambda = -1$ and $\lambda = -\frac{4}{5}$.

With $\lambda = -1$ we have from eqs. (4.18) and (4.20) that $a = c = 0$. Hence a solution is

$$x = 0 \qquad y = be^{-t} \qquad z = 0$$

With $\lambda = -\frac{4}{5}$ we have, from eqs. (4.19) and (4.20), that

$$a = -\frac{c(\frac{4}{5} + 1)}{(-\frac{4}{5} + 2)} = -\frac{3}{2}c \qquad b = -\frac{c(-\frac{4}{5} + 2)}{(-\frac{4}{5} + 1)} = -6c$$

Hence a solution is

$$x = -\tfrac{3}{2}ce^{-4t/5} \qquad y = -6ce^{-4t/5} \qquad z = ce^{-4t/5}$$

We note that the characteristic equation is of order two, although there are three equations. Hence the general solution has two arbitrary constants and is

$$x = -\tfrac{3}{2}ce^{-4t/5} \qquad y = be^{-t} - 6ce^{-4t/5} \qquad z = ce^{-4t/5}$$

Example 4.13 Determine the general solution to the pair of first-order differential equations

$$dx/dt + 2dy/dt - 3x + 4y = \alpha$$

$$2dx/dt + dy/dt + 2x - y = \beta$$

where (a) $\alpha = 1$, $\beta = 0$, (b) $\alpha = e^t$, $\beta = 0$

The general solution to the associated homogeneous differential equations has been obtained in Example 4.11; it therefore remains to obtain a particular solution.

(a) Try $x = a$, $y = b$.
On substituting into the system of differential equations, we obtain
$$-3a + 4b = 1 \qquad 2a - b = 0$$
Hence $a = \frac{1}{5}$, $b = \frac{2}{5}$, so that the general solution is
$$x = Ae^{-t/3} + Be^{-5t} + \tfrac{1}{5}$$
$$y = Ae^{-t/3} - \tfrac{4}{3}Be^{-5t} + \tfrac{2}{5}$$
(b) Try $x = ae^t$, $y = be^t$.
On substituting into the system of differential equations
$$(a + 2b - 3a + 4b)e^t = e^t$$
$$(2a + b + 2a - b)e^t = 0$$
Hence $a = 0$, $b = \frac{1}{6}$, so that the particular integral is $x = 0$, $y = \frac{1}{6}e^t$, and the general solution is the complementary function plus the particular integral.

Example 4.14 Find the charge on the capacitor of the circuit shown in Fig. 4.5, if the currents i_1 and i_2 and the charge on the capacitor are initially zero and if $E = 500 \sin 10t$ V, $R_1 = R_2 = 10$ ohms, $L = 1$ H, $C = 0.01$ F.

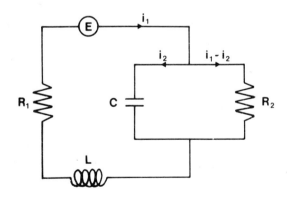

Fig. 4.5

The voltage drop across the resistance R_1 is $R_1 i_1 = 10\, dQ_1/dt$; across the resistance R_2 is $R_2(i_1 - i_2) = 10(dQ_1 - dQ_2)/dt$; across the induction L is $L\, di_1/dt = d^2 Q_1/dt^2$; across the capacitor C is $Q_2/C = 100 Q_2$; across the battery is $-E = -500 \sin 10t$.

From Kirchhoff's laws the differential equations governing the charge in the circuit are

$$d^2 Q_1/dt^2 + 10 dQ_1/dt + 100 Q_2 = 500 \sin 10t \qquad (4.21)$$

$$dQ_1/dt - dQ_2/dt - 10 Q_2 = 0 \qquad (4.22)$$

We consider the associated homogeneous system and try $Q_1 = a e^{\lambda t}$, $Q_2 = b e^{\lambda t}$. On substituting into the differential equations

$$a(\lambda^2 + 10\lambda) + 100 b = 0$$

$$a\lambda - b(\lambda + 10) = 0$$

The characteristic equation is therefore

$$(\lambda^2 + 10\lambda)(\lambda + 10) + 100\lambda = 0$$

with roots

$$\lambda = 0,\ \lambda = -10 \pm 10i$$

With

$$\lambda = 0,\ b = 0$$

with

$$\lambda = -10 + 10i,\ b = a(1 + i)$$

with

$$\lambda = -10 - 10i,\ b = a(1 - i)$$

Hence the general solution to the associated homogeneous system is

$$Q_1 = a + e^{-10t}(b e^{10it} + c e^{-10it})$$

$$Q_2 = e^{-10t}[b(1+i) e^{10it} + c(1-i) e^{-10it}]$$

On putting $\exp(\pm 10it) = \cos 10t \pm i \sin 10t$ this solution can be rewritten as

$$Q_1 = A + e^{-10t}[B \cos 10t + C \sin 10t]$$

$$Q_2 = e^{-10t}[(B + C) \cos 10t + (C - B) \sin 10t]$$

For the particular solution try

$$Q_1 = a\cos 10t + b\sin 10t \qquad Q_2 = c\cos 10t + d\sin 10t$$

On substituting into the pair of differential equations (4.21) and (4.22) and equating coefficients of $\sin 10t$ and $\cos 10t$, four equations are obtained, with solution $a = -3$, $b = -1$, $c = -2$, $d = 1$. Therefore the general solution is

$$Q_1 = A + e^{-10t}(B\cos 10t + C\sin 10t) - (3\cos 10t + \sin 10t)$$
$$Q_2 = e^{-10t}[(B + C)\cos 10t + (C - B)\sin 10t] - (2\cos 10t - \sin 10t)$$

The given initial conditions are

$$Q_2(0) = 0 \qquad Q_2'(0) = 0 \qquad Q_1'(0) = 0$$

but we note that there are in effect only two independent initial conditions, since from eq. (4.22), Q_1' must be zero if Q_2 and Q_2' are both zero. Hence two equations result for the constants B and C:

$$B + C - 2 = 0$$
$$-10(B + C) + 10(C - B) + 10 = 0$$

with solution $B = \frac{1}{2}$, $C = \frac{3}{2}$.

Thus the charge on the capacitor at time t is

$$Q_2(t) = \sin 10t - 2\cos 10t + e^{-10t}(\sin 10t + 2\cos 10t)$$

The solution of systems of homogeneous differential equations of the type considered are of great importance in many scientific and engineering problems. For example, the motion of many physical systems can be represented by such sets of equations. These equations can then be rewritten in matrix notation as

$$Ay = \lambda y \qquad (4.23)$$

where for the given system A is a known square matrix, and where the solution of the system of differential equations can be expressed in terms of the *eigenvalues* λ and the *eigenvectors* y. Powerful numerical methods exist for the determination of the eigenvalues and eigenvectors of any matrix.

Example 4.15 (Students unfamiliar with eigenvalues and eigenvectors should omit this example and Problem 4.8.)

Analytic Solution of Second- and Higher-Order Differential Equations

Determine the motion, about the static equilibrium position, of the two spring system of Section 1.2.2 with $(\lambda/Lm_1) = 2$ and $(\lambda/Lm_2) = 3$.

Let x_1, x_2 be the displacement of the masses m_1, m_2 from their equilibrium position; then the differential equations (1.10) become

$$d^2x_1/dt^2 = -4x_1 + 2x_2$$
$$d^2x_2/dt^2 = 3x_1 - 3x_2$$
(4.24)

where

$$z_1 = x_1 + \tfrac{5}{6}g + \tfrac{1}{3}(\lambda/m_2)$$
$$z_2 = x_2 + \tfrac{7}{6}g + \tfrac{2}{3}(\lambda/m_2)$$

In matrix notation, eqs. (4.24) can be written as

$$\ddot{x} = Ax \qquad (4.25)$$

where

$$x = \begin{bmatrix} x_1 \\ x_2 \end{bmatrix} \qquad \ddot{x} = \begin{bmatrix} d^2x_1/dt^2 \\ d^2x_2/dt^2 \end{bmatrix} \qquad A = \begin{bmatrix} -4 & 2 \\ 3 & -3 \end{bmatrix}$$

As usual we try a solution

$$x = ye^{\omega t}$$

where y is a column matrix with constants as elements. Differentiating and substituting into the matrix equation (4.25), we obtain

$$Ay = \lambda y$$

where $\lambda = \omega^2$.

This matrix is of the form of eq. (4.23). By the methods of matrix algebra, the eigenvalues and eigenvectors of the matrix A are found to be

$$\lambda = -1 \qquad y_{(-1)} = \begin{bmatrix} 2 \\ 3 \end{bmatrix}$$

$$\lambda = -6 \qquad y_{(-6)} = \begin{bmatrix} 1 \\ -1 \end{bmatrix}$$

When $\lambda = -1$, $\omega = \pm i$ and when $\lambda = -6$, $\omega = \pm\sqrt{6}i$, so the general solution to eq. (4.24) is

$$x = ay_{(-1)}e^{it} + by_{(-1)}e^{-it} + cy_{(-6)}e^{\sqrt{6}it} + dy_{(-6)}e^{-\sqrt{6}it}$$

108 Ordinary Differential Equations

This can be written as

$$x = y_{(-1)}(E \cos t + B \sin t) + y_{(-6)}(C \cos \sqrt{6}t + D \sin \sqrt{6}t)$$

Hence

$$x_1 = 2(E \cos t + B \sin t) + (C \cos \sqrt{6}t + D \sin \sqrt{6}t)$$
$$x_2 = 3(E \cos t + B \sin t) - (C \cos \sqrt{6}t + D \sin \sqrt{6}t)$$

Problem 4.7 Find the general solutions to the following systems of differential equations

(a) $2\,dx/dt + dy/dt - 2x = 6(2t + 1)$
    $2\,dx/dt + dy/dt + x\ \ = 3t$

(b) $dx/dt - dy/dt - y = 0$
    $dy/dt + x - y\ \ \ = 0$

(c) $2\,dx/dt + dy/dt - 2x + 3y = -1$
    $2\,dx/dt + dy/dt + 4x + y\ \ = t$

(d) $dx/dt + dy/dt + 2x + y = 2 + e^t$
    $dy/dt + dz/dt + y - z\ \ = 3 + e^t$
    $dx/dt + dz/dt + 2x - z = 1 + e^t$

Problem 4.8 Find the eigenvalues and eigenvectors of the matrix

$$\begin{bmatrix} -2 & -1 \\ 4 & 3 \end{bmatrix}$$

Hence find the general solution to the pair of simultaneous differential equations

$$dx/dt = -2x - y$$
$$dy/dt = 4x + 3y$$

Problem 4.9 Find the solution to the pair of equations

$$d^2x/dt^2 = 0 \qquad d^2y/dt^2 = -g$$

subject to the initial conditions $x(0) = y(0) = 0$,

$$x'(0) = V \cos \alpha \qquad y'(0) = V \sin \alpha$$

Hence find the relationship between y and x.

4.6 Other second- and higher-order equations

4.6.1 Dependent variable absent

If the differential equation is of the form

$$f\left(\frac{d^n x}{dt^n}, \frac{d^{n-1} x}{dt^{n-1}}, \ldots, \frac{dx}{dt}, t\right) = 0$$

the substitution $dx/dt = p$, $d^2x/dt^2 = dp/dt$, etc, reduces the order of the differential equation by one.

Example 4.16 Find the general solution to

$$2\frac{d^2 x}{dt^2} - \left(\frac{dx}{dt}\right)^2 + 1 = 0$$

Let $dx/dt = p$; hence $d^2x/dt^2 = dp/dt$, so that $2\, dp/dt = p^2 - 1$. This is a separable first-order differential equation (see Section 2.2) with solution

$$\ln\left(\frac{p-1}{p+1}\right) = t + A$$

Hence $(p-1)/(p+1) = ae^t$, or

$$\frac{dx}{dt} = p = \frac{1 + ae^t}{1 - ae^t}$$

This is again a first-order separable differential equation, and so

$$x(t) = b + \int \frac{1 + ae^t}{1 - ae^t}\, dt = b + t - 2\ln(1 - ae^t)$$

is the general solution of the original equation.

4.6.2 Independent variable absent

If the differential equation is of the form

$$f\left(\frac{d^n x}{dt^n}, \frac{d^{n-1} x}{dt^{n-1}}, \ldots, \frac{dx}{dt}, x\right) = 0$$

the substitution $dx/dt = p$, $d^2x/dt^2 = (dx/dt)\, d/dx(dx/dt) = p\, dp/dx$, etc. reduces the order of the differential equation by one.

Example 4.17 Find the general solution to

$$x\frac{d^2x}{dt^2} + \left(\frac{dx}{dt}\right)^2 = 1$$

Let $dx/dt = p$; hence $d^2x/dt^2 = p\, dp/dx$, so that $xp\, dp/dx = 1 - p^2$. This is a separable first-order differential equation with solution

$$p^2 = 1 - A/x^2$$

Hence

$$\frac{dx}{dt} = \pm \frac{\sqrt{(x^2 - A)}}{x}$$

This is again a first-order separable differential equation with solution

$$t + B = \pm\sqrt{(x^2 - A)}$$

After squaring both sides, the solution can be written as

$$x^2 = t^2 + bt + a$$

4.6.3 Euler's linear equation

These equations are of the form

$$t^n\frac{d^nx}{dt^n} + a_1 t^{n-1}\frac{d^{n-1}x}{dt^{n-1}} + \cdots + a_{n-1} t\frac{dx}{dt} + a_n x = f(t) \qquad (4.26)$$

where all the a's are real constants. Equation (4.26) can be transformed into a linear equation with constant coefficients, considered in Sections 4.2 and 4.3, by the transformation $t = e^z$.

Clearly $dt/dz = e^z$, and hence

$$\frac{dx}{dt} = \frac{dz}{dt}\frac{dx}{dz} = e^{-z}\frac{dx}{dz}$$

and

$$\frac{d^2x}{dt^2} = \frac{dz}{dt}\frac{d}{dz}\left(\frac{dx}{dt}\right) = e^{-z}\frac{d}{dz}\left(e^{-z}\frac{dx}{dz}\right) = e^{-2z}\left(\frac{d^2x}{dz^2} - \frac{dx}{dz}\right)$$

$$= e^{-2z}\frac{d}{dz}\left(\frac{d}{dz} - 1\right)x$$

Analytic Solution of Second- and Higher-Order Differential Equations   111

In general

$$\frac{d^n x}{dt^n} = e^{-nz} \frac{d}{dz}\left(\frac{d}{dz} - 1\right)\left(\frac{d}{dz} - 2\right) \cdots \left(\frac{d}{dz} - n + 1\right) x$$

When these expressions are substituted into the differential equation, the transformed differential equation has constant coefficients.

Example 4.18   Find the general solution to

$$t^3\, d^3x/dt^3 + 3t^2\, d^2x/dt^2 - 2t\, dx/dt + 2x = \ln t$$

Let $t = e^z$ so that $\ln t = z$. On substituting the expressions for the derivatives into the differential equation, we obtain

$$\frac{d}{dz}\left(\frac{d}{dz} - 1\right)\left(\frac{d}{dz} - 2\right)x + 3\frac{d}{dz}\left(\frac{d}{dz} - 1\right)x - 2\frac{dx}{dz} + 2x = z$$

or after simplification, $d^3x/dz^3 - 3\, dx/dz + 2x = z$.

The characteristic equation is $m^3 - 3m + 2 = 0$, with roots 1, 1, -2. For the particular integral try $x = az + b$, which leads to the general solution

$$x = Ae^z + Bze^z + Ce^{-2z} + \tfrac{1}{4}(2z + 3)$$

and on transforming to the original variable t,

$$x = At + Bt \ln t + C/t^2 + \tfrac{1}{4}(2 \ln t + 3)$$

Problem 4.10   Find the general solution to the following differential equations:

(a) $\dfrac{d^2x}{dt^2} - \left(\dfrac{dx}{dt}\right)^2 + 1 = 0$

(b) $\left(\dfrac{d^3x}{dt^3}\right)^2 + t\dfrac{d^3x}{dt^3} - \dfrac{d^2x}{dt^2} = 0$

(c) $x\dfrac{d^2x}{dt^2} = 2\left(\dfrac{dx}{dt}\right)^2 - \left(\dfrac{dx}{dt}\right)$

(d) $t^3 \dfrac{d^3x}{dt^3} + 2t\dfrac{dx}{dt} - 2x = t$

Miscellaneous problems

Problem 4.11 Find the general solutions to the following differential equations:

(a) $x'' + 4x' + 13x = t + 4 \sin t$

(b) $x''' - 6x'' + 12x' - 8x = te^{2t}$

(c) $2x' + 3y' + z'' + 2z' + z = 1$

$$y + z' = 0$$
$$x' + y' - z = 0$$

(d) $t^2 x'' - tx' + x = 3t + 1$

CHAPTER FIVE

Numerical Solutions of Second- and Higher-Order Differential Equations

5.1 Introductory discussion

The numerical solution of a first-order differential equation must by its very nature be an initial-value problem. With a given condition at the value of the independent variable $x = x_a$, the solution can, by marching forwards, be obtained for values of $x > x_a$ and, by marching backwards, be obtained for $x < x_a$. If we are concerned with initial-value problems for differential equations of any order, then by methods which are simple extensions of those discussed in Chapter 3, the solution can be obtained by marching forwards or backwards. If, on the other hand, the problem is a boundary-value one, then we do not have enough conditions at a single point to be able to march either forwards or backwards. As a result of this the solution of a boundary-value problem is, in most cases, much more time consuming than that of an initial-value problem.

The comments made in Section 3.5 about the ill-conditioning and stability of first-order equations carry over directly to higher-order equations. However, for these equations there is the possibility that the equations are so-called 'stiff' differential equations. The numerical solution of stiff equations raises further difficulties which are discussed in Section 5.4.

5.2 Initial-value problems

5.2.1 Series solution

Solution in series is considered in Chapter 6. If, as in this case, the problem is an initial-value one, then the solution can be obtained as a Taylor series as described in Section 6.2, or the general solution in series can be obtained as described in Sections 6.3 and 6.4, and the constants evaluated on using the initial conditions. Having obtained the series solution, the numerical solution can be obtained to any desired accuracy for any value of the independent variable within the radius of convergence of the series. If a large number of terms are needed to obtain the solution to the required accuracy, it is better to use a number of steps. For example, suppose that the initial conditions are given at $x = a$ and the solution is required at $x = c$. With the given conditions, the solution in series about $x = a$, that is a series involving powers of $x - a$, can be obtained. On putting $x = c$, the solution can be evaluated. However, the solution can also be obtained by finding the solution at an intermediate point $x = b$ and hence the solution in series about $x = b$ which is then used to find the required solution at $x = c$. Fewer terms in the series are needed in the computation of the solutions at $x = b$ and $x = c$, but two series have had to be evaluated.

Example 5.1 Find y and y' when $x = 0 \cdot 2$, given that

$$(1 + x^2)y'' + xy' - y = 0$$

subject to $y(0) = 1$ and $y'(0) = 0$.

The series solution is found in Example 6.1 to be

$$y = 1 + \tfrac{1}{2}x^2 - \frac{3}{4!}x^4 + \frac{5 \cdot 3^2 \cdot 1}{6!}x^6 - \frac{7 \cdot 5^2 \cdot 3^2 \cdot 1}{8!}x^8 + \ldots$$

and hence

$$y' = x - \frac{1}{2!}x^3 + \frac{3^2 \cdot 1}{4!}x^5 - \frac{5^2 \cdot 3^2 \cdot 1}{6!}x^7 + \ldots$$

both series being convergent for $|x| \leq 1$.
Substituting $x = 0 \cdot 2$ into the series we obtain the solution correct to four decimal places as

$$y(0 \cdot 2) = 1 \cdot 0198 \qquad y'(0 \cdot 2) = 0 \cdot 1961$$

5.2.2 Runge–Kutta formulae

Runge–Kutta formulae can be obtained for second- and higher-order differential equations and for systems of differential equations in a manner very similar to that described for first-order equations in Section 3.3.2. Formulae are listed for two types of second-order equations and for a pair of first-order equations, although it should be remembered that an nth-order differential equation can be written as a set of n first-order equations.

A *third-order* formula for

$$y'' = f(x, y)$$

is

$$\bar{y}_1 = y_0 + h[y'_0 + \tfrac{1}{6}(k_1 + 2k_2)]$$
$$\bar{y}'_1 = y'_0 + \tfrac{1}{6}(k_1 + 4k_2 + k_3) \tag{5.1}$$

where

$$k_1 = hf(x_0, y_0)$$
$$k_2 = hf(x_0 + \tfrac{1}{2}h, y_0 + \tfrac{1}{2}hy'_0 + \tfrac{1}{8}hk_1)$$
$$k_3 = hf(x_0 + h, y_0 + hy'_0 + \tfrac{1}{2}hk_2)$$

The local errors in $\bar{y}_1$ and $\bar{y}'_1$ are proportional to h^4.

A *fourth-order* formula for

$$y'' = f(x, y, y')$$

is

$$\bar{y}_1 = y_0 + h[y'_0 + \tfrac{1}{6}(k_1 + k_2 + k_3)]$$
$$\bar{y}'_1 = y'_0 + \tfrac{1}{6}(k_1 + 2k_2 + 2k_3 + k_4) \tag{5.2}$$

where

$$k_1 = hf(x_0, y_0, y'_0)$$
$$k_2 = hf(x_0 + \tfrac{1}{2}h, y_0 + \tfrac{1}{2}hy'_0 + \tfrac{1}{8}hk_1, y'_0 + \tfrac{1}{2}k_1)$$
$$k_3 = hf(x_0 + \tfrac{1}{2}h, y_0 + \tfrac{1}{2}hy'_0 + \tfrac{1}{8}hk_1, y'_0 + \tfrac{1}{2}k_2)$$
$$k_4 = hf(x_0 + h, y_0 + hy'_0 + \tfrac{1}{2}hk_3, y'_0 + k_3)$$

The local error is proportional to h^5.

A *fourth-order* formula for

$$y' = f(x, y, z) \qquad z' = g(x, y, z)$$

is

$$\bar{y}_1 = y_0 + \tfrac{1}{6}(k_1 + 2k_2 + 2k_3 + k_4) \qquad (5.3)$$
$$\bar{z}_1 = z_0 + \tfrac{1}{6}(l_1 + 2l_2 + 2l_3 + l_4)$$

where

$k_1 = hf(x_0, y_0, z_0);$ $\qquad l_1 = hg(x_0, y_0, z_0)$
$k_2 = hf(x_0 + \tfrac{1}{2}h, y_0 + \tfrac{1}{2}k_1, z_0 + \tfrac{1}{2}l_1);$ $\; l_2 = hg(x_0 + \tfrac{1}{2}h, y_0 + \tfrac{1}{2}k_1, z_0 + \tfrac{1}{2}l_1)$
$k_3 = hf(x_0 + \tfrac{1}{2}h, y_0 + \tfrac{1}{2}k_2, z_0 + \tfrac{1}{2}l_2);$ $\; l_3 = hg(x_0 + \tfrac{1}{2}h, y_0 + \tfrac{1}{2}k_2, z_0 + \tfrac{1}{2}l_2)$
$k_4 = hf(x_0 + h, y_0 + k_3, z_0 + l_3);$ $\qquad l_4 = hg(x_0 + h, y_0 + k_3, z_0 + l_3)$

The local error is proportional to h^5.

It should be noted how similar the fourth-order formula, (5.3), for a pair of first-order equations is to the equivalent formula, (3.13), for a single first-order equation. The extension to a system of n first-order equations follows in the obvious manner. An estimate of the error in the solution can also be obtained in a similar manner to that for a single first-order equation.

Example 5.2   Find the solution to Example 5.1 using a fourth-order Runge–Kutta formula.

We could use (5.2) with

$$f(x, y, y') = \frac{y - xy'}{1 + x^2}$$

However, we will consider a pair of first-order equations and use (5.3). Let $y' = z$, so that the second-order equation can be written as

$$y' = z \qquad z' = \frac{y - xz}{1 + x^2}$$

with $y(0) = 1$ and $z(0) = 0$. Hence

$$f(x, y, z) = z \qquad g(x, y, z) = (y - xz)/(1 + x^2)$$

(a) Two steps of equal size.

First step $x_0 = 0, y_0 = 1, z_0 = 0, h = 0{\cdot}1$

Using (5.3),

$k_1 = 0{\cdot}1 f(0, 1, 0) = 0; \, l_1 = 0{\cdot}1 g(0, 1, 0) = 0{\cdot}1$

$k_2 = 0.1f(0.05, 1, 0.05) = 0.005; l_2 = 0.1g(0.05, 1, 0.05) = 0.099\ 50$
$k_3 = 0.1f(0.05, 1.0025, 0.049\ 75) = 0.004\ 98;$
$\qquad l_3 = 0.1g(0.05, 1.0025, 0.049\ 75) = 0.099\ 75$
$k_4 = 0.1f(0.1, 1.004\ 98, 0.099\ 75) = 0.009\ 98\cdot$
$\qquad l_4 = 0.1g(0.1, 1.004\ 98, 0.099\ 75) = 0.098\ 52$

Hence $\bar{y}(0.1) = 1.004\ 99, \bar{z}(0.1) = 0.099\ 50.$

$\quad$ Second step $x_1 = 0.1, \bar{y}_1 = 1.004\ 99, \bar{z}_1 = 099\ 50, h = 0.1$
$k_1 = 0.1f(0.1, 1.004\ 99, 0.099\ 50) = 0.009\ 95;$
$\qquad l_1 = 0.1g(0.1, 1.004\ 99, 0.099\ 50) = 0.098\ 52$
$k_2 = 0.1f(0.15, 1.009\ 97, 0.148\ 76) = 0.014\ 88;$
$\qquad l_2 = 0.1g(0.15, 1.009\ 97, 0.148\ 76) = 0.096\ 59$
$k_3 = 0.1f(0.15, 1.012\ 43, 0.147\ 80) = 0.014\ 78;$
$\qquad l_3 = 0.1g(0.15, 1.012\ 43, 0.147\ 80) = 0.096\ 85$
$k_4 = 0.1f(0.2, 1.019\ 77, 0.196\ 35) = 0.019\ 64;$
$\qquad l_4 = 0.1g(0.2, 1.019\ 77, 0.196\ 35) = 0.094\ 28$

Hence to four decimal places
$$\bar{y}(0.2) = 1.0198 \qquad \bar{z}(0.2) = 0.1961$$

$\quad$ (b) One step: $x_0 = 1, y_0 = 1, z_0 = 0, h = 0.2.$
$k_1 = 0.2f(0, 1, 0) = 0; l_1 = 0.2g(0, 1, 0) = 0.2$
$k_2 = 0.2f(0.1, 1, 0.1) = 0.02; l_2 = 0.2g(0.1, 1, 0.1) = 0.196\ 04$
$k_3 = 0.2f(0.1, 1.01, 0.098\ 02) = 0.019\ 60;$
$\qquad l_3 = 0.2g(0.1, 1.01, 0.098\ 02) = 0.198\ 06$
$k_4 = 0.2f(0.2, 1.019\ 60, 0.198\ 06) = 0.039\ 61;$
$\qquad l_4 = 0.2g(0.2, 1.019\ 60, 0.198\ 06) = 0.188\ 46$
Hence to four decimal places
$$\bar{y}^*(0.2) = 1.0198 \qquad \bar{z}^*(0.2) = 0.1961$$

Since the error is proportional to h^5, and since the values from two and one steps in (a) and (b) are the same to four decimal places, we are reasonably confident that the solutions as given will be correct to this number of decimal places, and we see that they agree with the values from the series solution in Example 5.1.

Problem 5.1 By expansion in series and comparing with the Taylor series for $y(x)$, verify the formulae (5.1)–(5.3).

Problem 5.2 Find y and y' at $x = 0\cdot4$ with (a) two steps of equal size, and (b) one step, using the third-order Runge–Kutta formula (5.1), given that
$$y'' = x - 4y \qquad \text{with } y(0) = 1, \ y'(0) = \tfrac{1}{4}$$
Use the two values to obtain an estimate of their accuracy and compare against the numerical value obtained from the analytic solution.

Problem 5.3 Find y and y' at $x = 0\cdot2$ in one step, using the fourth-order Runge–Kutta formula (a) (5.2), (b) (5.3), given that
$$y'' + y' + y = x \qquad \text{with } y(0) = 0, \ y'(0) = 0\cdot5$$

5.2.3 Predictor–corrector formulae

The predictor–corrector methods extend directly to second- and higher-order equations. We consider, by way of example, the second-order equation
$$y'' = f(x, y, y') \qquad (5.4)$$
and assume that y and y' are known at values of x up to and including x_0.

The predicted value for y' at $x = x_1$, where $x_1 - x_0 = h$, can be obtained by use of (3.22), and is
$$\bar{z}_1^p = z_{-3} + \tfrac{4}{3}h(2f_0 - f_{-1} + 2f_{-2}) \qquad (5.5)$$
where for ease of presentation we have let $y' = z$. Since we now have an estimate for y_1', namely $\bar{z}_1^p$, we can use (3.24) to obtain an estimate for y_1. That is
$$\bar{y}_1^{c(1)} = y_{-1} + \tfrac{1}{3}h[\bar{z}_1^p + 4z_0 + z_{-1}] \qquad (5.6)$$
We note that it has not been necessary to use a predictor formula for y_1, since an estimate of y_1' has already been obtained.

We now iterate to obtain
$$\bar{z}_1^{c(1)} = z_{-1} + \tfrac{1}{3}h[f(x_1, \bar{y}_1^{c(1)}, \bar{z}_1^p) + 4f_0 + f_{-1}]$$
$$\bar{y}_1^{c(2)} = y_{-1} + \tfrac{1}{3}h[\bar{z}_1^{c(1)} + 4z_0 + z_{-1}]$$
$$\bar{z}_1^{c(2)} = z_{-1} + \tfrac{1}{3}h[f(x_1, \bar{y}_1^{c(2)}, \bar{z}_1^{c(1)}) + 4f_0 + f_{-1}]$$
$$\vdots$$

Numerical Solutions of Second- and Higher-Order Differential Equations   119

When the process has converged, the approximate error in $\bar{z}_1^c$ is given by $-\frac{1}{29}(\bar{z}_1^c - \bar{z}_1^p)$.

We next consider a pair of first-order equations:

$$y' = f(x, y, z) \qquad z' = g(x, y, z) \qquad (5.7)$$

We obtain a predicted value for y and z at $x = x_1$ using each of the equations, then using the predicted values, obtain corrected values for both y and z.

Hence

$\bar{y}_1^p = y_{-3} + \frac{4}{3}h(2f_0 - f_{-1} + 2f_{-2}); \bar{z}_1^p = z_{-3} + \frac{4}{3}h(2g_0 - g_{-1} + 2g_{-2})$

$\bar{y}_1^{c(1)} = y_{-1} + \frac{1}{3}h(f_1^p + 4f_0 + f_{-1}); \quad \bar{z}_1^{c(1)} = z_{-1} + \frac{1}{3}h(g_1^p + 4g_0 + g_{-1})$

$\bar{y}_1^{c(2)} = y_{-1} + \frac{1}{3}h(f_1^{c(1)} + 4f_0 + f_{-1}); \bar{z}_1^{c(2)} = z_{-1} + \frac{1}{3}h(g_1^{c(1)} + 4g_0 + g_{-1})$

$\vdots$

where

$$f_1^{c(r)} = f[x_1, \bar{y}_1^{c(r-1)}, \bar{z}_1^{c(r-1)}]$$

and similarly for $g_1^{c(r)}$. A measure of the error in $\bar{y}_1^c$ is again given by

$$-\tfrac{1}{29}(\bar{y}_1^c - \bar{y}_1^p)$$

Example 5.3 In one step find y and y' when $x = 1.4$ given

$$x^2 y'' - xy' + y = 1$$

together with the values in Table 5.1.

Table 5.1

x	1	1·1	1·2	1·3
y	1	1·104 841	1·218 786	1·341 074
$z = y'$	1	1·095 310	1·182 322	1·262 364

(a) With $x_0 = 1.3$, $x_{-1} = 1.2$, $x_{-2} = 1.1$, $x_{-3} = 1$, $h = 0.1$ and $f(x, y, z) = (1 - y + xz)/x^2$
then $f_0 = 0.769\,230$, $f_{-1} = 0.833\,333$, $f_{-2} = 0.909\,091$.
Hence

$\bar{z}_1^p = z_{-3} + \frac{4}{3}h(2f_0 - f_{-1} + 2f_{-2}) = 1.336\,441$

$\bar{y}_1^{c(1)} = y_{-1} + \frac{1}{3}h(\bar{z}_1^p + 4z_0 + z_{-1}) = 1.471\,060$

$$f(1\cdot4, 1\cdot471\,060, 1\cdot336\,441) = 0\cdot714\,264$$

$$\bar{z}_1^{c(1)} = z_{-1} + \tfrac{1}{3}h(0\cdot714\,264 + 4f_0 + f_{-1}) = 1\cdot336\,473$$

$$\bar{y}_1^{c(2)} = y_{-1} + \tfrac{1}{3}h(\bar{z}_1^{c(1)} + 4z_0 + z_{-1}) = 1\cdot471\,061$$

$$f(1\cdot4, 1\cdot471\,061, 1\cdot336\,473) = 0\cdot714\,286$$

$$\bar{z}_1^{c(2)} = z_{-1} + \tfrac{1}{3}h(0\cdot714\,286 + 4f_0 + f_{-1}) = 1\cdot336\,473$$

Hence $\bar{z}_1^c = 1\cdot336\,473$ and $\bar{y}_1^c = 1\cdot471\,061$. A measure of the accuracy in $\bar{z}_1$ is given by

$$-\tfrac{1}{29}(\bar{z}_1^c - \bar{z}_1^p) = -1\cdot1 \times 10^{-6}$$

Note again that $\bar{y}_1^{c(1)}$ and $\bar{z}_1^{c(1)}$ can be taken as the estimates for y_1 and z_1.

(b) With

$$y' = z[= f(x, y, z)]; \quad z' = (1 - y + xz)/x^2 \; [= g(x, y, z)]$$

so that

$f_0 = 1\cdot262\,364 \qquad f_{-1} = 1\cdot182\,322 \qquad f_{-2} = 1\cdot095\,310 \qquad f_{-3} = 1$

and

$g_0 = 0\cdot769\,230 \qquad g_{-1} = 0\cdot833\,333 \qquad g_{-2} = 0\cdot909\,091 \qquad g_{-3} = 1$

we obtain

$\bar{y}_1^p = 1\cdot471\,070; \; \bar{z}_1^p = 1\cdot336\,441$

$f_1^p = f(1\cdot4, 1\cdot471\,070, 1\cdot336\,441) = 1\cdot336\,441;$
$\qquad\qquad g_1^p = g(1\cdot4, 1\cdot471\,070, 1\cdot336\,441) = 0\cdot714\,259$

$\bar{y}_1^{c(1)} = 1\cdot471\,060; \; \bar{z}_1^{c(1)} = 1\cdot336\,472$

$f_1^{c(1)} = f(1\cdot4, 1\cdot471\,060, 1\cdot336\,472) = 1\cdot336\,472;$
$\qquad\qquad g_1^{c(1)} = g(1\cdot4, 1\cdot471\,060, 1\cdot336\,472) = 0\cdot714\,286$

$\bar{y}_1^{c(2)} = 1\cdot471\,061; \; \bar{z}_1^{c(2)} = 1\cdot336\,473$

Another iteration gives the same values, so that

$$\bar{y}_1^c = 1\cdot471\,061 \qquad \bar{z}_1^c = 1\cdot336\,473$$

which are the same values as obtained in (a).

Numerical Solutions of Second- and Higher-Order Differential Equations

Problem 5.4 In one step find y and y' when $x = 0.8$, given

$$y'' + y' + y = x$$

together with the values in Table 5.2.

Table 5.2

x	0	0.2	0.4	0.6
y	0	0.091 299	0.170 096	0.243 038
y'	0.5	0.419 303	0.374 216	0.359 913

5.3 Boundary-value problems

We consider second-order differential equations with given values of the dependent variable at two different values of the independent variable. The methods of solution which are described can however be applied or developed for other differential equations and/or boundary conditions.

5.3.1 Series solution

We take the boundary conditions to be

$$y = y_a \text{ at } x = x_a \text{ and } y = y_b \text{ at } x = x_b \tag{5.8}$$

and assume that the general expansion in series has been obtained by methods described in Chapter 6. The series solution involves two constants which can be evaluated on using the boundary conditions (5.8) providing that x_a and x_b lie within the radius of convergence of the series. The solution can then be evaluated for any value of x within that radius.

Example 5.4 Find the solution at $x = 1.5$, correct to four decimal places, of

$$x^2 y'' + xy' + (x^2 - \tfrac{1}{4})y = 0$$

subject to $y(1) = 0$ and $y(2) = 1$.

The general solution in series given in Example 6.8 is

$$y = Ax^{1/2} \sum_{r=0}^{\infty} \frac{(-1)^r}{(2r+1)!} x^{2r} + Bx^{-1/2} \sum_{r=0}^{\infty} \frac{(-1)^r}{(2r)!} x^{2r}$$

which is convergent for $|x| > 0$.

Using the boundary conditions, we obtain

$$0 = A\left[1 - \frac{1}{3!} + \frac{1}{5!} - \cdots\right] + B\left[1 - \frac{1}{2!} + \frac{1}{4!} - \cdots\right]$$

$$1 = A\sqrt{2}\left[1 - \frac{2^2}{3!} + \frac{2^4}{5!} - \cdots\right] + \frac{B}{\sqrt{2}}\left[1 - \frac{2^2}{2!} + \frac{2^4}{4!} - \cdots\right]$$

On evaluating the series to five decimal places and solving for A and B, we obtain the values $A = 0{\cdot}908\,06$ and $B = -1{\cdot}414\,22$. Using these values together with $x = 1{\cdot}5$, the required solution is $y(1{\cdot}5) = 0{\cdot}6579$ to four decimal places.

5.3.2 Trial-and-error Method

A value of the derivative y'_a is guessed. Then using the given value y_a and the guessed value y'_a, a solution of the differential equation can be obtained at x_b using a marching process involving Runge–Kutta and predictor–corrector formulae. The resulting value of y at x_b is not likely to be the required value. Another guess for y'_a is made and the corresponding solution at x_b is obtained. Using the two values at x_b, a better estimate of y'_a can be made. The process is repeated until the pair of values taken for the initial-value problem gives the required value y_b at x_b. The method involves a lot of computation, but will eventually succeed if the problem is well-conditioned to the marching process.

Example 5.5 Determine $y'(0)$ given that

$$y'' + y = x$$

with $y(0) = 1$ and $y(0{\cdot}5) = 1{\cdot}8$.

We note that the analytic solution is

$$y = 0{\cdot}881\,09 \sin x + \cos x + x$$

We will obtain the value of y at $x = 0{\cdot}5$ from given initial conditions at $x = 0$, using 5 equal steps and the third-order Runge–Kutta formula (5.1). For two guessed values of $y'(0)$, namely $y'(0) = 0$ and $y'(0) = 1$, the resulting values of $\bar{y}$ and $\bar{y}'$ are given in Table 5.3.

Table 5.3

x	0	0·1	0·2	0·3	0·4	0·5
$\bar{y}$	1	0·995 17	0·981 40	0·959 82	0·931 65	0·898 16
$\bar{y}'$	0	−0·094 84	−0·178 74	−0·250 86	−0·310 49	−0·357 02
$\bar{y}$	1	1·095 00	1·180 06	1·255 33	1·321 05	1·377 57
$\bar{y}'$	1	0·900 17	0·801 33	0·704 48	0·610 58	0·520 57

Linear extrapolation with the values at $x = 0.5$ to obtain a value $y(0.5) = 1.8$ gives a value for $y'(0)$ as follows:

$$\frac{1 \cdot 8 - 1 \cdot 377\ 57}{1 \cdot 377\ 57 - 0 \cdot 898\ 16} = \frac{y'(0) - 1}{1 - 0}$$

whence $y'(0) = 1.881\ 14$.

We now try the initial conditions $y(0) = 1$, $y'(0) = 1.881\ 14$; the resulting solution is as shown in Table 5.4.

Table 5.4

$\bar{y}$	1	1·182 97	1·355 12	1·515 73	1·664 19	1·800 02
$\bar{y}'$	1·881 14	1·776 91	1·664 91	1·546 27	1·422 17	1·293 85

The initial values $y(0) = 1.0$ and $y'(0) = 1.881\ 14$ have given the boundary condition at $x = 0.5$ correct to four decimal places. The solution obtained with these initial conditions are therefore the required solution. The reason that linear extrapolation from two guessed values gave the required value for the initial condition is that the differential equation is linear and of second order. If the equation is linear and of nth order, then n different sets of initial conditions are needed to obtain the correct initial value. If the equation is non-linear, then repeated calculations are needed to converge to the required solution. However, since the solution at x_b will be a continuous function of the initial values, various methods can be developed which speed up the convergence of the process.

Problem 5.5 Find $y'(0)$, using a fourth-order Runge–Kutta formula with one step, given

$$y'' + y' + y = x \quad \text{with } y(0) = 0, y(0\cdot 2) = 0\cdot 5$$

5.3.3 Algebraic solution

The differential equation is changed into a set of algebraic equations on approximating the derivatives in terms of solution values. We illustrate the method by solving again the differential equation of Example 5.5.

We again take five steps (see Fig. 5.1).

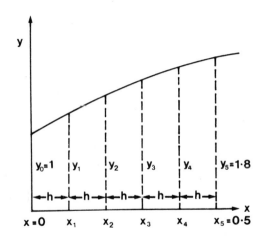

Fig. 5.1

The values y_0 and y_5 are known to be 1 and 1·8 respectively.

The derivative, y'_1, at x_1 will be approximated by the slope of a straight line passing through the points $(x_{1/2}, y_{1/2})$ and $(x_{3/2}, y_{3/2})$. That is

$$\left(\frac{dy}{dx}\right)_1 \simeq \frac{y_{3/2} - y_{1/2}}{x_{3/2} - x_{1/2}} = \frac{y_{3/2} - y_{1/2}}{h}$$

Numerical Solutions of Second- and Higher-Order Differential Equations

Similarly
$$y'_2 \simeq (y_{5/2} - y_{3/2})/h, \quad y'_{1/2} \simeq (y_1 - y_0)/h, \text{ etc.}$$

In a similar way we approximate the second derivatives:

$$\left(\frac{d^2y}{dx^2}\right)_1 \simeq \frac{y'_{3/2} - y'_{1/2}}{h} \simeq \frac{(y_2 - y_1) - (y_1 - y_0)}{h^2} = \frac{y_2 - 2y_1 + y_0}{h^2}$$

and
$$y''_2 \simeq (y_3 - 2y_2 + y_0)/h^2, \text{ etc.}$$

The approximation for y''_1 together with the values y_1 and x_1 are substituted into the differential equation, $y'' + y = x$, to obtain

$$(y_2 - 2y_1 + y_0)/h^2 + y_1 = x_1$$

Three other equations are obtained in the same way at x_2, x_3, x_4. The four equations can be rewritten as

$$\begin{aligned}
-(2-h^2)y_1 + y_2 &= -y_0 + h^2 x_1 \\
y_1 - (2-h^2)y_2 + y_3 &= h^2 x_2 \\
y_2 - (2-h^2)y_3 + y_4 &= h^2 x_3 \\
y_3 - (2-h^2)y_4 &= -y_5 + h^2 x_4
\end{aligned}$$
(5.9)

The simultaneous equations (5.9) involve the four unknowns y_1, y_2, y_3, y_4 and can be solved in various ways to give, with $y_0 = 1$, $y_5 = 1 \cdot 8$, $h = 0 \cdot 1$,

$y_1 = 1 \cdot 182\ 99 \qquad y_2 = 1 \cdot 355\ 14 \qquad y_3 = 1 \cdot 515\ 75 \qquad y_4 = 1 \cdot 664\ 20$

The differential equation which has been considered did not involve the derivative y'. For differential equations which do, it is necessary to approximate y' in terms of solution values, that is in terms of y_0, y_1, y_2, etc. Hence we take

$$y'_1 \simeq \frac{y_2 - y_0}{2h} \qquad y'_2 \simeq \frac{y_3 - y_1}{2h} \quad \text{etc.}$$

If the differential equation is linear, then the resulting set of simultaneous equations for y_1 are also linear; however if the differential equation is non-linear, then the simultaneous equations for y_1 are non-linear. The solution of a non-linear set of equations is no easy matter, and the student is referred to specialist books on the subject. We note again that the solution of a non-linear differential equation can be difficult and time consuming.

Problem 5.6 Find $y(0 \cdot 1)$, given

$$y'' + y' + y = x \qquad \text{with } y(0) = 0, \ y(0 \cdot 2) = 0 \cdot 5$$

5.4 Stiff equations

A differential equation which has exponential solutions, $e^{\lambda x}$, where the λ have widely differing values and where the smallest value of λ is a fairly large negative number, is known as a *stiff equation*.
Some examples of stiff equations are

$$y'' + 21y' + 20y = 0 \qquad (5.10)$$
$$y'' - 100y = 0 \qquad (5.11)$$
$$y'' - L(x)y = 0 \qquad L(x) \gg 0 \qquad (5.12)$$
$$y'' + ay' - L(x,y)y = 0 \qquad L(x,y) \gg 0 \qquad (5.13)$$

We have already noted in Section 3.5.2 that if Euler's method is applied to the differential equation $y' = -10y$, which has a solution $y = Ae^{-10x}$, then the step length has to be less than a critical value, otherwise the numerical solution becomes unstable. A similar result holds for any Runge–Kutta formula applied to the equation $y' = -10y$, or indeed any equation resulting in a general solution involving a rapidly decreasing exponential. The critical step size is related to the largest negative exponential appearing in the general solution, even though this exponential does not appear in the particular solution with given initial or boundary conditions. For example, a numerical solution obtained by a fourth-order Runge–Kutta method becomes unstable for a step length h if $|h\lambda_L| > 2 \cdot 8$ (approximately), where λ_L is the largest negative value of the λ. The result is illustrated by obtaining the solution of (5.10) subject to the condition $y(0) = 1$, $y'(0) = -1$, and using a fourth-order Runge–Kutta method. The values of λ for this differential equation are $-1, -20$ so that $\lambda_L = -20$.

The analytic solution subject to the initial conditions is $y = e^{-x}$. Two step lengths are taken: (a) $h = 0 \cdot 1$, (b) $h = 0 \cdot 2$. For (a),

$|h\lambda_L| = 2$, and for (b), $|h\lambda_L| = 4$, so that we expect the numerical solution (b) to become unstable. The numerical solutions for integral values of x are given in Table 5.5 and compared against values obtained from the analytic solution.

Table 5.5

x	$h = 0 \cdot 1$	$h = 0 \cdot 2$	analytic
1·0	0·367 88	0·367 88	0·367 88
2·0	0·135 34	0·135 34	0·135 34
3·0	0·049 79	0·037 71	0·049 79
4·0	0·018 32	−37·727 98	0·018 32
5·0	0·006 74	−117 957·168 03	0·006 74

If the general solution of the stiff equation has an increasing exponential as well as a rapidly decreasing exponential, as have (5.11)–(5.13), then in most physical situations it is the negative exponential which is required. This leads to further difficulties which can be illustrated by considering the solution of (5.11). The general solution is $y = Ae^{-10x} + Be^{10x}$, but we assume that because of physical considerations the required solution is $y = Ae^{-10x}$. However any error in the numerical solution has the effect of introducing the unwanted increasing exponential, so that the solution obtained is of the form $y = Ae^{-10x} + \epsilon e^{10x}$, and the contribution from the increasing exponential does not remain small for long, even though ϵ might be very small (see Figs. 5.2 and 5.3).

Note that this instability will occur even though the step size is smaller than the critical value required for any Runge–Kutta method. The difficulties arising from the unwanted increasing solution can be reduced as follows:

(a) For some stiff equations the *Riccati transformation*

$$y' = \eta(x)y \tag{5.14}$$

can be useful. For example, using the transformation, (5.12) becomes

$$\eta' + \eta^2 = L(x) \tag{5.15}$$

which is a non-linear first-order equation.
Having obtained η from (5.15), the solution y is given by (5.14) as

$$\ln y = \int \eta(x)\,dx$$

which can be evaluated using methods given in Section 3.2.

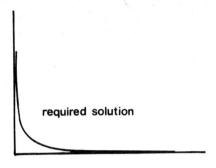

Fig. 5.2

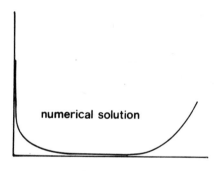

Fig. 5.3

(b) If the Riccati transformation is not helpful, then we can reverse the direction of integration. That is, we start the integration procedure at a large value of x and find the solution for smaller values of x. In this way the required solution is the increasing exponential, and the unwanted solution is a decreasing exponential which then gets smaller as the solution process progresses. Unfortunately, the starting values are now not known, so that a trial-and-error process is required.

Stiff equations have another difficulty associated with their numerical solution in that any error in y leads to a very much bigger error in y'', which is needed in any predictor–corrector cycle. The subsequent integration of y'' decreases this error by an amount dependent upon the step size, so that by taking small enough steps, the error introduced by terms of the type Ly can be kept within bounds. However, this is time consuming and it is better to reverse the predictor–corrector cycle. That is, an estimate of y_1 is obtained by an extrapolation formula. Using this estimate a numerical differentiation formula can be used to estimate y_1''. Then from the differential equation a better estimate of y_1 can be obtained and the cycle repeated. Numerical differentiation increases any error, but since y_1 is obtained from y_1'' by effectively dividing by L, the errors are kept small.

CHAPTER SIX

Series solution

6.1 Introduction

In many cases, the solution of a differential equation may be required in a small neighbourhood of some given point. If the solution of the differential equation cannot be obtained in closed form by methods described in the previous chapters, then the desired behaviour of the solution may be obtained from a series solution.

In this chapter we consider series solutions of differential equations of the form

$$P_0(x)\frac{d^2y}{dx^2} + P_1(x)\frac{dy}{dx} + P_2(x)y = Q(x) \qquad (6.1)$$

These differential equations are linear, and in cases in which $Q(x) \equiv 0$, they are also homogeneous. It should be borne in mind that if y_1 and y_2 are independent solutions of the homogeneous differential equations, then other solutions are Ay_1 and By_2 and the general solution is $(Ay_1 + By_2)$.

6.2 Solution by Maclaurin, Taylor series

Consider the second-order differential equation with given initial conditions $y(0) = y_0$ and $y^{(1)}(0) = y_1$. Then if the solution $y(x)$ can be expressed as a Maclaurin expansion, the series solution is

$$y(x) = y(0) + xy^{(1)}(0) + \frac{x^2}{2!}y^{(2)}(0) + \cdots + \frac{x^n}{n!}y^{(n)}(0) + \cdots \quad (6.2)$$

where

$$y^{(n)} = d^n y/dx^n$$

Now the values of $y(0)$ and $y^{(1)}(0)$ are known, and the value of $y^{(2)}(0)$ can be found from the differential equation with $x = 0$, that is,

$$y^{(2)}(0) = \frac{Q(0) - P_2(0)y_0 - P_1(0)y_1}{P_0(0)}$$

By differentiating the differential equation (6.1) and putting $x = 0$, we can obtain the value of $y^{(3)}(0)$. This process can be repeated to obtain the higher derivatives, and therefore the Maclaurin expansion (6.2) for $y(x)$ can be found.

For this method to work, we see from the expression for $y^{(2)}(0)$ that a necessary condition is $P_0(0) \neq 0$. When this condition is satisfied, $x = 0$ is called an *ordinary point* of the differential equation. When $P_0(0) = 0$, then $x = 0$ is called a *singular point* of the differential equation. It is also obvious that the series solution is valid only within its radius of convergence, and further that the series is of practical use only if a small number of terms in the series are sufficient to give the solution to the required numerical accuracy.

Example 6.1 Find the series solution of

$$(1 + x^2)y'' + xy' - y = 0$$

subject to

$$y(0) = 1 \qquad y'(0) = 0$$

In Table 6.1, we list the differential equation and its derivatives and those equations with $x = 0$. From these equations $y^{(n)}$ can be obtained. Hence the series solution for $y(x)$ up to and including the term involving x is

$$y(x) = 1 + \tfrac{1}{2}x^2 - \frac{3}{4!}x^4 + \cdots$$

Table 6.1

Differential equation and its derivatives	Differential equation and its derivatives with $x = 0$	$y^{(n)}(0)$
$(1+x^2)y^{(2)} + xy^{(1)} - y = 0$ $(1+x^2)y^{(3)} + 2xy^{(2)} + xy^{(2)} + y^{(1)} - y^{(1)}$ $\quad = (1+x^2)y^{(3)} + 3xy^{(2)} = 0$ $(1+x^2)y^{(4)} + 2xy^{(3)} + 3xy^{(3)} + 3y^{(2)}$ $\quad = (1+x^2)y^{(4)} + 5xy^{(3)} + 3y^{(2)} = 0$	$y^{(2)}(0) - y(0) = 0$ $y^{(3)}(0) = 0$ $y^{(4)}(0) + 3y^{(2)}(0) = 0$	$\left.\begin{array}{l} y(0) = 1 \\ y^{(1)}(0) = 0 \\ y^{(2)}(0) = 1 \end{array}\right\}$ given $y^{(3)}(0) = 0$ $y^{(4)}(0) = -3$

If only a few terms in the series are required, then it is sufficient to differentiate the differential equation as shown. However, if more terms are needed, and if in particular the general term in the series is required so that convergence can be investigated then it is desirable to use *Leibnitz's theorem*, which states that if u and v are functions of x, then

$$\frac{d^n}{dx^n}(uv) = u^{(n)}v + \binom{n}{1}u^{(n-1)}v^{(1)} + \cdots + \binom{n}{r}u^{(n-r)}v^{(r)} + \cdots + uv^{(n)}$$

If we apply this to $(1+x^2)y^{(2)}$ with $u = y^{(2)}$ and $v = (1+x^2)$, then

$$\frac{d^n}{dx^n}[(1+x^2)y^{(2)}] = y^{(n+2)}(1+x^2) + \binom{n}{1}y^{(n+1)}2x + \binom{n}{2}y^{(n)}2$$

$$= (1+x^2)y^{(n+2)} + 2nx\, y^{(n+1)} + n(n-1)y^n$$

since

$$\frac{d^n(1+x^2)}{dx^n} = 0 \qquad \text{for } n \geqslant 3$$

Similarly

$$\frac{d^n}{dx^n}(xy^{(1)}) = xy^{(n+1)} + ny^{(n)}$$

The nth derivative of the given differential equation is therefore
$$(1 + x^2)y^{(n+2)} + (2n + 1)xy^{(n+1)} + (n^2 - 1)y^{(n)} = 0$$
and hence with $x = 0$
$$y^{(n+2)}(0) = -(n^2 - 1)y^{(n)}(0)$$
This recurrence relation enables us to build up the value of $y^{(n+2)}(0)$ as follows:

$$\left.\begin{array}{l} y(0) = 1 \\ y^{(1)}(0) = 0 \end{array}\right\} \text{given}$$

$y^{(2)}(0) = 1$ from given differential equation

$n = 1, y^{(3)}(0) = 0$

$n = 2, y^{(4)}(0) = -3 \cdot 1 y^{(2)}(0) = -3 \cdot 1$

$n = 3, y^{(5)}(0) = 0$ since $y^{(3)}(0) = 0$

$n = 4, y^{(6)}(0) = -5 \cdot 3 y^{(4)}(0) = 5 \cdot 3^2 \cdot 1$

It follows that $y^{(n)}(0) = 0$ when n is odd. For n even,
$$y^{(n+2)}(0) = (-1)^{n/2}(n+1)(n-1)^2(n-3)^2 \cdots 3^2 \cdot 1$$
The solution is up to and including the term involving x^8,
$$y(x) = 1 + \tfrac{1}{2}x^2 - \frac{3}{4!}x^4 + \frac{5 \cdot 3^2 \cdot 1}{6!}x^6 - \frac{7 \cdot 5^2 \cdot 3^2 \cdot 1}{8!}x^8 + \cdots$$
The coefficient of the general term x^n, for n even, is given by
$$\frac{y^{(n)}(0)}{n!} = (-1)^{n/2-1} \frac{(n-1)(n-3)^2 \cdots 3^2 \cdot 1}{n(n-1)(n-2)(n-3) \cdots 3 \cdot 2 \cdot 1}$$
$$= (-1)^{n/2-1} \frac{(n-3)(n-5) \cdots 3 \cdot 1}{2^{n/2}(\tfrac{1}{2}n)!}$$
$$= (-1)^{n/2-1} \frac{(n-3)!}{2^{n/2}(\tfrac{1}{2}n)! 2^{(n-4)/2}(\tfrac{1}{2}n - 2)!}$$

Since the series involves only even powers of x, we replace n by $2r$, so that r can take all positive integral values and
$$y(x) = 1 + \tfrac{1}{2}x^2 + \sum_{r=2}^{\infty}(-1)^{r-1}\frac{(2r-3)!}{2^{2r-2}r!(r-2)!}x^{2r}$$

Series Solution 133

Using the ratio test for convergence (Barnett and Cronin (1975), Section 1.5.1), the series is convergent if

$$\lim_{r \to \infty} \left| \frac{(2r-1)! \, x^{2r+2}}{2^{2r}(r+1)!(r-1)!} \cdot \frac{2^{2r-2} r!(r-2)!}{(2r-3)! \, x^{2r}} \right| < 1$$

that is,

$$\lim_{r \to \infty} \left| \frac{(2r-1)(2r-2)x^2}{4(r+1)(r-1)} \right| < 1$$

that is,

$$|x^2| < 1$$

Hence the series is convergent for $-1 < x < 1$.

The values $x = \pm 1$ must be investigated separately. In both cases the series are of alternating type with the terms tending to zero as $r \to \infty$. Hence (Barnett and Cronin (1975), Section 1.5.2) the series is convergent for both $x = 1$ and $x = -1$, so the series is convergent for $-1 \leq x \leq 1$, and divergent for $|x| > 1$.

If a solution of the differential equation is required in the neighbourhood of $x = a$, then provided that $x = a$ is an ordinary point of the differential equation, that is $P_0(a) \neq 0$, a solution is obtained in terms of a Taylor expansion,

$$y(x) = y(a) + y^{(1)}(a)(x-a) + \frac{y^{(2)}(a)}{2!}(x-a)^2 + \cdots$$

However, when evaluating the series, it is better to transform the independent variable from x to z such that the point $x = a$ corresponds to $z = 0$. That is, we let $z = x - a$ and obtain a Maclaurin expansion in z.

The method outlined above can be extended to differential equations of order other than second, as illustrated in the following example.

Example 6.2 Find the series solution, up to and including the term involving x^4, of the differential equation

$$yy' + y = x^2$$

subject to the initial condition $y(0) = 1$.

Table 6.2 shows the required derivatives, their values at $x = 0$, and the calculated values of $y^{(n)}(0)$.

Table 6.2

Differential equation and its derivatives	Differential equation and its derivatives with $x = 0$	$y^{(n)}(0)$
$yy^{(1)} + y = x^2$ $yy^{(2)} + (y^{(1)})^2 + y^{(1)} = 2x$ $yy^{(3)} + 3y^{(1)}y^{(2)} + y^{(2)} = 2$ $yy^{(4)} + 4y^{(1)}y^{(3)} + 3(y^{(2)})^2 + y^{(3)} = 0$	$y^{(1)}(0) + 1 = 0$ $y^{(2)}(0) + 1 - 1 = 0$ $y^{(3)}(0) = 2$ $y^{(4)}(0) - 8 + 2 = 0$	$y(0) = 1$ given $y^{(1)}(0) = -1$ $y^{(2)}(0) = 0$ $y^{(3)}(0) = 2$ $y^{(4)}(0) = 6$

From these,

$$y = 1 - x + \frac{x^3}{3} + \frac{x^4}{4} + \cdots$$

Problem 6.1 By obtaining the solution of the following differential equations with given initial conditions by the methods given in Chapter 4, and by solution in series, find the series expansions of $\sin x$ and $e^{-x} \cos x$.

(a) $y'' + y = 0$, $\quad y(0) = 0, \quad y'(0) = 1$

(b) $y'' + 2y' + 2y = 0$, $\quad y(0) = 1, \quad y'(0) = -1$

Problem 6.2 Using Maclaurin expansions, find the solutions in series of the following differential equations with the given initial conditions, and state the convergence of the series:

(a) $y'' - xy' + y = 0$, $\quad y(0) = y'(0) = 1$

(b) $(1 - x^2)y'' - 2xy' + \frac{3}{4}y = 0$, $\quad y(0) = 1, y'(0) = 0$.

Problem 6.3 Find the value of y and of y', correct to four decimal places, at $x = 0 \cdot 1$ for the solutions obtained for Problems 6.1 and 6.2.

6.3 General solution in series about an ordinary point

In this section we consider solutions in series about an ordinary point of a second-order differential equation, when the initial conditions are *not* given. That is, we again consider the differential equation (6.1) where $P_0(a) \neq 0$, and obtain a solution in series about $x = a$, of the form

$$y(x) = A_0 + A_1(x - a) + A_2(x - a)^2 + \cdots$$

where the constant coefficients A_n are to be determined.

It will be assumed that the functions $P_0(x), P_1(x), P_2(x), Q(x)$ are either polynomials in x or can be expanded in Taylor series about $x = a$.

The method of solution is to substitute the series for y, y' and y'' into the differential equation and then to equate coefficients of $(x - a)^n$. A series of equations for A_n are obtained, from which all the coefficients can be determined in terms of any two of the coefficients. These two coefficients are then the two arbitrary constants expected in the general solution of a second-order equation, and can be determined if any two conditions are given.

Example 6.3 Find a solution in series about $x = 0$ of the differential equation

$$(1 - x^2)y'' - 2xy' + 6y = 0$$

Since $P_0(x) \equiv 1 - x^2$ and $P_0(0) = 1$, $x = 0$ is an ordinary point of the equation and a solution of form

$$y = A_0 + A_1 x + A_2 x^2 + \cdots + A_n x^n + \cdots$$

can be obtained.
Differentiating, we obtain

$$y' = A_1 + 2A_2 x + \cdots + nA_n x^{n-1} + \cdots$$

and

$$y'' = 2A_2 + \cdots + n(n-1)A_n x^{n-2} + \cdots$$

It is convenient to list the coefficients of x^n for each of the terms $y'', -x^2 y'', -2xy', 6y$ in tabular form (Table 6.3).

Series Solution   137

Table 6.3

Term	Constant	x	x^2	x^n
y''	$2A_2$	$3 \cdot 2 A_3$	$4 \cdot 3 A_4$	$(n+2)(n+1)A_{n+2}$
$-x^2 y''$			$-2A_2$	$-n(n-1)A_n$
$-2xy'$		$-2A_1$	$-2 \cdot 2 A_2$	$-2nA_n$
$6y$	$6A_0$	$6A_1$	$6A_2$	$6A_n$

Note that, for this example, the constant and x terms are special, in that not all the terms in the differential equation contribute to these coefficients. It is wise to check the general term x^n by taking the lowest value of n that is not a special case. In this case if we put $n = 2$ into the general values we obtain, running down the column,

$$4 \cdot 3 A_4 \quad -2A_2 \quad -2 \cdot 2 A_2 \quad 6A_2$$

which check with coefficients of x^2 listed.

We now sum each column in the table and equate against the corresponding coefficient on the right-hand side of the equation, which in this case is zero.
Hence

$$2A_2 + 6A_0 = 0 \qquad A_2 = -3A_0$$

$$6A_3 + 4A_1 = 0 \qquad A_3 = -\tfrac{2}{3} A_1$$

$$12 A_4 = 0 \qquad A_4 = 0$$

$$(n+2)(n+1)A_{n+2} - (n+3)(n-2)A_n = 0 \qquad A_{n+2} = \frac{(n+3)(n-2)}{(n+2)(n+1)} A_n$$

From these, the general series solution is

$$y = A_0(1 - 3x^2)$$

$$+ A_1 \left(x - \frac{2}{3} x^3 - \frac{1}{5} x^5 - \frac{4}{7 \cdot 5} x^7 - \cdots \right.$$

$$\left. - \frac{(n+1)}{(2n+1)(2n-1)} x^{2n+1} - \cdots \right)$$

and the series is convergent for $-1 < x < 1$.

Example 6.4 Find the general solution in series about $x = 0$ of the equations

(a) $2x(1 - x)y'' + (1 - 5x)y' - y = 0$
(b) $x^2 y'' + xy + (x^2 - \frac{1}{4})y = 0$

We note that $x = 0$ is not an ordinary point of either differential equation (but is a singular point). However, we will attempt to obtain a series solution of the form

$$y = \sum_{n=0}^{\infty} A_n x^n$$

(a) The resulting coefficients of powers of x are as shown in Table 6.4.

Table 6.4

Term	Constant	x	x^2	x^n
$2xy''$		$2 \cdot 2A_2$	$2 \cdot 3 \cdot 2A_3$	$2 \cdot (n+1)n A_{n+1}$
$-2x^2 y''$			$-2 \cdot 2A_2$	$-2n(n-1)A_n$
y'	A_1	$2A_2$	$3A_3$	$(n+1)A_{n+1}$
$-5xy'$		$-5A_1$	$-5 \cdot 2A_2$	$-5nA_n$
$-y$	$-A_0$	$-A_1$	$-A_2$	$-A_n$

Equating coefficients, we have

$$\begin{array}{lll} \text{constant} & A_1 = A_0 & \\ x & 6A_2 = 6A_1 & \left. \begin{array}{l} A_0 = A_1 = A_2 = \cdots \\ = A_n = \cdots \end{array} \right. \\ x^2 & 15A_3 = 15A_2 & \\ x^n & (2n+1)(n+1)A_{n+1} = (2n+1)(n+1)A_n & \end{array}$$

It follows that a solution is

$$y = A_0(1 + x + x^2 + \cdots + x^n + \cdots)$$

This series is convergent for $-1 < x < 1$ and then sums to $(1-x)^{-1}$, so that a solution is $y = A_0(1-x)^{-1}$. This cannot be the general solution to the differential equation since it involves only one arbitrary constant. The series associated with the other constant is not therefore of the form

$$\sum_{r=0}^{\infty} A_r x^r$$

(b) The resulting coefficients of powers of x are as shown in Table 6.5. On equating coefficients we see that

$$0 = A_0 = A_1 = A_2 = \cdots = A_n = \cdots$$

Hence for this differential equation there is no series solution of the form $\sum_{r=0}^{\infty} A_r x^r$.

The form of the series solution valid about a singular point is discussed in the next section.

Table 6.5

Term	Constant	x	x^2	x^n
$x^2 y''$			$2A_2$	$n(n-1)A_n$
xy'		A_1	$2A_2$	nA_n
$x^2 y$			A_0	A_{n-2}
$-\tfrac{1}{4}y$	$-\tfrac{1}{4}A_0$	$-\tfrac{1}{4}A_1$	$-\tfrac{1}{4}A_2$	$-\tfrac{1}{4}A_n$

Problem 6.4 Find the general solutions in series of the differential equations given in Problem 6.2, and check the solutions against those obtained in the problem for the given initial conditions.

Problem 6.5 Find the general solution in series, valid for large values of x, of the differential equation

$$x^2(1 + x^2)y'' + x(1 + 2x^2)y' - y = 0$$

Find the radius of convergence. (Hint: let $z = 1/x$ so that $dy/dx = -z^2 \, dy/dz$ and $d^2y/dx^2 = z^4 \, d^2y/dz^2 + 2z^3 \, dy/dz$.)

6.4 General solution in series about a regular singular point

We have noted that $x = a$ is a singular point of the differential equation

$$P_0(x)\frac{d^2y}{dx^2} + P_1(x)\frac{dy}{dx} + P_2(x)y = 0 \qquad (6.3)$$

if $P_0(a) = 0$.
In this case, let

$$P_0(x) = (x - a)R_1(x) = (x - a)^2 R_2(x)$$

Then (6.3) can be rewritten as

$$\frac{d^2y}{dx^2} + \frac{1}{(x-a)}\left[\frac{P_1(x)}{R_1(x)}\right]\frac{dy}{dx} + \frac{1}{(x-a)^2}\left[\frac{P_2(x)}{R_2(x)}\right]y = 0 \qquad (6.4)$$

$x = a$ is a *regular singular point* of the differential equation if

$$P_0(a) = 0 \qquad \lim_{x \to a}\frac{P_1(x)}{R_1(x)} = \alpha \qquad \lim_{x \to a}\frac{P_2(x)}{R_2(x)} = \beta$$

where α and β are finite.

Example 6.5 Find the singular points, and determine their nature for the differential equations

(a) $2x(1-x)y'' + (1-5x)y' - y = 0$
(b) $x^2 y'' + xy' + (x^2 - \frac{1}{4})y = 0$
(c) $x^3 y'' + x^2 y' + (x^2 - \frac{1}{4})y = 0$
(d) $x(x^2 + 1)y'' + y' + y = 0$

(a) Singular points are given by $2x(1-x) = 0$, that is, $x = 0$ and $x = 1$.
For $x = 0$, rewrite the equation as

$$\frac{d^2y}{dx^2} + \frac{1}{x}\frac{1-5x}{2(1-x)}\frac{dy}{dx} + \frac{1}{x^2}\frac{-x}{2(1-x)}y = 0$$

Then

$$\lim_{x \to 0}\frac{1-5x}{2(1-x)} = \frac{1}{2} \qquad \lim_{x \to 0}\frac{-x}{2(1-x)} = 0$$

and since both limits are finite, $x = 0$ is a regular singular point.

For $x = 1$, rewrite the equation as

$$\frac{d^2y}{dx^2} + \frac{1}{(1-x)} \frac{1-5x}{2x} \frac{dy}{dx} + \frac{1}{(1-x)^2} \frac{-(1-x)}{2x} y = 0$$

$$\lim_{x \to 1} \frac{1-5x}{2x} = -2 \qquad \lim_{x \to 1} \frac{-(1-x)}{2x} = 0$$

so that $x = 1$ is a regular singular point.

(b) $x = 0$ is a singular point.
Rewrite the equation as

$$\frac{d^2y}{dx^2} + \frac{1}{x} [1] \frac{dy}{dx} + \frac{1}{x^2} (x^2 - \tfrac{1}{4}) y = 0$$

$$\lim_{x \to 0} (1) = 1 \qquad \lim_{x \to 0} (x^2 - \tfrac{1}{4}) = -\tfrac{1}{4}$$

so that $x = 0$ is a regular singular point.

(c) $x = 0$ is a singular point.
Rewrite the equation as

$$\frac{d^2y}{dx^2} + \frac{1}{x} (1) \frac{dy}{dx} + \frac{1}{x^2} \left(x - \frac{1}{4x}\right) y = 0$$

$$\lim_{x \to 0} 1 = 1 \qquad \lim_{x \to 0} \left(x - \frac{1}{4x}\right) \to \pm \infty$$

In this case, one of the limits does not exist, so that $x = 0$ is a singular point but not a regular singular point.

(d) Singular points are given by $x(x^2 + 1) = 0$; that is $x = 0$, $x = +i$, $x = -i$, where $i = \sqrt{(-1)}$. Hence there is one real singular point and two imaginary singular points. The real singular point is easily shown to be a regular singular point.

We are concerned only with functions of a real variable, and will consider solution in series about the regular singular point $x = 0$, since if a solution in series about another regular singular point is required, the equation can be transformed as described in Section 6.2. If $x = 0$ is a regular singular point, then a solution in series exists of the form

$$y = x^m (A_0 + A_1 x + \cdots + A_n x^n + \cdots) \qquad (6.5)$$

where $A_0 = 1$ and the index m is to be determined.

A theorem states that the above series always converges in the region of the complex plane bounded by two circles centred at $x = 0$. The radius of one circle is arbitrarily small while the other passes through the nearest singular point. Further, if all the singular points are at $x = 0$, then the series is convergent for $|x| > 0$. The convergence of the series at $x = 0$ has to be considered as a special case. When $P_0(x)$ has singular points other than $x = 0$, and is or can be written as a product of real linear and quadratic functions in x, the theorem takes a special simplified form.

Let

$$P_0(x) = Ax^k(x - \alpha_1)(x - \alpha_2) \cdots (x - \alpha_n)(x^2 + 2a_1 x + b_1)$$
$$\times (x^2 + 2a_2 x + b_2) \cdots (x^2 + 2a_m x + b_m)$$

where, since we assume that the quadratics cannot be written as a product of two real linear functions, $b_r > a_r^2$. Using the theorem, it is easily shown that the radius of convergence of the series is given by the smallest value of

$$|\alpha_1|, |\alpha_2|, \ldots, |\alpha_n|, \sqrt{b_1}, \ldots, \sqrt{b_m}$$

Example 6.6 Determine the region of convergence of the series about the regular singular point, $x = 0$, of the differential equations of Example 6.5 and of the equation

$$x(2x + 1)(x^2 + x + 1)y'' + y' + y = 0$$

(a) $P_0(x) \equiv -2x(x - 1)$ hence $\alpha_1 = 1$ and the region of convergence given by $0 < |x| < 1$.
(b) There is a singular point only at $x = 0$, so that the series is convergent for all x except possibly $x = 0$, that is the series is convergent for $|x| > 0$.
(d) $P_0(x) \equiv x(x^2 + 1)$; hence $b_1 = 1$ and the series is convergent for $0 < |x| < 1$.
(e) $x = 0$ is a regular singular point with $P_0(x) = 2x(x + \tfrac{1}{2})(x^2 + x + 1)$; hence $|\alpha_1| = \tfrac{1}{2}$, $b = 1$ and the region of convergence is $0 < |x| < \tfrac{1}{2}$.

The method of solution starts in a similar way to that for ordinary points, namely the series for y, y' and y'' are substituted into the equation and the coefficients of powers of x equated to zero. A set of equations then result for the coefficients A_n and the index m.

Example 6.7 Find the general solution of the differential equation

$$2x(1-x)y'' + (1-5x)y' - y = 0$$

This is the differential equation previously considered in Examples 6.4(a), 6.5(a) and 6.6(a).

We have seen in these examples that $x = 0$ is a regular singular point; hence, let

$$y = x^m \sum_{r=0}^{\infty} A_r x^r \qquad A_0 = 1$$

that is,

$$y = A_0 x^m + A_1 x^{m+1} + A_2 x^{m+2} + \cdots + A_n x^{m+n} + \cdots$$

Hence

$$y' = mA_0 x^{m-1} + (m+1)A_1 x^m + (m+2)A_2 x^{m+1} + \cdots$$

$$+ (m+n)A_n x^{m+n-1} + \cdots$$

and

$$y'' = m(m-1)A_0 x^{m-2} + (m+1)mA_1 x^{m-1} + (m+2)(m+1)A_2 x^m + \cdots$$

$$+ (m+n)(m+n-1)A_n x^{m+n-2} + \cdots$$

Setting up the table of coefficients of powers of x and noting that the lowest coefficient is x^{m-1}, we obtain Table 6.6. The general term should again be checked by, say, putting $n = 1$ and comparing against the x^{m+1} values.

Table 6.6

Term	x^{m-1}	x^m	x^{m+1}	x^{m+n}
$2xy''$	$2m(m-1)A_0$	$2(m+1)mA_1$	$2(m+2)(m+1)A_2$	$2(m+n+1)(m+n)A_{n+1}$
$-2x^2 y''$		$-2m(m-1)A_0$	$-2(m+1)mA_1$	$-2(m+n)(m+n-1)A_n$
y'	mA_0	$(m+1)A_1$	$(m+2)A_2$	$(m+n+1)A_{n+1}$
$-5xy'$		$-5mA_0$	$-5(m+1)A_1$	$-5(m+n)A_n$
$-y$		$-A_0$	$-A_1$	$-A_n$

Equating the sums of the columns to zero, we obtain

$$m(2m - 1)A_0 = 0$$
$$(m + 1)(2m + 1)A_1 - (m + 1)(2m + 1)A_0 = 0$$
$$(m + 2)(2m + 3)A_2 - (m + 2)(2m + 3)A_1 = 0$$
$$\vdots$$
$$(m + n + 1)(2m + 2n + 1)A_{n+1} - (m + n + 1)(2m + 2n + 1)A_n = 0$$

since $A_0 = 1$ we have that

$$m(2m - 1) = 0$$

and hence either $m = 0$ or $m = \frac{1}{2}$. The equation for m is called the *indicial equation*.

The other equations are satisfied by

$$1 = A_0 = A_1 = A_2 = \cdots = A_n = \cdots$$

regardless of the value of m. It follows that with $m = \frac{1}{2}$ a solution is given by

$$y_1 = x^{1/2}(1 + x + x^2 + \cdots + x^n + \cdots)$$

Similarly it follows that with $m = 0$ a solution is given by

$$y_2 = (1 + x + x^2 + \cdots x^n + \cdots)$$

which is independent of the solution y_1. But the differential equation is a homogeneous linear differential equation, and hence another solution is

$$y = A(1 + x + x^2 + \cdots x^n + \cdots) + Bx^{1/2}(1 + x + x^2 + \cdots + x^n + \cdots)$$

which is the general solution since it involves two arbitrary constants. It is to be noted that the first series is that obtained in Example 6.4(a), and that the region of convergence given in Example 6.6(a) is $0 < |x| < 1$. However, the series is obviously convergent for $x = 0$; hence

$$y = (A + Bx^{1/2}) \sum_{r=0}^{\infty} x^r$$

is convergent for $-1 < x < 1$.

The series can be summed within its radius of convergence to give the solution

$$y = \frac{A + Bx^{1/2}}{1 - x}$$

For this example, the coefficients A_n were independent of the value of m. In general this is not the case, as illustrated in the next example.

Example 6.8 Find the general solution in series about $x = 0$ of
$$x^2 y'' + xy' + (x^2 - \tfrac{1}{4})y = 0$$

$x = 0$ is a regular singular point of the differential equation, so that a solution of the form
$$y = x^m \sum_{r=0}^{\infty} A_r x^r$$
exists. The coefficients of powers of x are as shown in Table 6.7.

Table 6.7

Term	x^m	x^{m+1}	x^{m+2}	x^{m+n}
$x^2 y''$	$m(m-1)A_0$	$(m+1)mA_1$	$(m+2)(m+1)A_2$	$(m+n)(m+n-1)A_n$
xy'	mA_0	$(m+1)A_1$	$(m+2)A_2$	$(m+n)A_n$
$x^2 y$			A_0	A_{n-2}
$-\tfrac{1}{4}y$	$-\tfrac{1}{4}A_0$	$-\tfrac{1}{4}A_1$	$-\tfrac{1}{4}A_2$	$-\tfrac{1}{4}A_n$

Equating the sums of the columns to zero gives
$$(m^2 - \tfrac{1}{4})A_0 = 0$$
$$[(m+1)^2 - \tfrac{1}{4}]A_1 = 0$$
$$[(m+2)^2 - \tfrac{1}{4}]A_2 + A_0 = 0$$
$$[(m+n)^2 - \tfrac{1}{4}]A_n + A_{n-2} = 0$$

Hence $m = \pm\tfrac{1}{2}$, and
$$A_1 = 0$$
$$A_2 = -\frac{A_0}{(m+\tfrac{5}{2})(m+\tfrac{3}{2})}$$
$$A_n = -\frac{A_{n-2}}{(m+n+\tfrac{1}{2})(m+n-\tfrac{1}{2})}$$

Using the recurrence relation and noting that $A_1 = 0$, it follows that all the odd indexed coefficients are zero.

We now obtain the even indexed coefficients by taking the two values of m in turn.

(a) $m = \tfrac{1}{2}$

$$A_2 = -\frac{A_0}{3 \cdot 2} \qquad A_n = -\frac{A_{n-2}}{(n+1)n}$$

Hence

$$A_4 = -\frac{A_2}{5 \cdot 4} = +\frac{A_0}{5 \cdot 4 \cdot 3 \cdot 2} = \frac{A_0}{5!}$$

$$A_6 = -\frac{A_4}{7 \cdot 6} = -\frac{A_0}{7 \cdot 6 \cdot 5!} = -\frac{A_0}{7!}$$

and

$$A_n = (-1)^{n/2} A_0 / (n+1)! \qquad n \text{ even}$$

Hence a solution to the differential equation is

$$y_1 = x^{1/2}\left(1 - \frac{x^2}{3!} + \frac{x^4}{5!} - \cdots\right) = x^{1/2} \sum_{r=0}^{\infty} \frac{(-1)^r}{(2r+1)!} x^{2r}$$

(b) $m = -\tfrac{1}{2}$

$$A_2 = -\frac{A_0}{2 \cdot 1} \qquad A_n = -\frac{A_{n-2}}{n(n-1)}$$

Hence

$$A_4 = -\frac{A_2}{4 \cdot 3} = +\frac{A_0}{4 \cdot 3 \cdot 2 \cdot 1} = \frac{A_0}{4!}$$

$$A_6 = -\frac{A_4}{6 \cdot 5} = -\frac{A_0}{6 \cdot 5 \cdot 4!} = -\frac{A_0}{6!}$$

and in general

$$A_n = (-1)^{n/2} A_0 / n! \qquad n \text{ even}$$

Hence a solution to the differential equation is

$$y_2 = x^{-1/2}\left(1 - \frac{x^2}{2!} + \frac{x^4}{4!} - \cdots\right) = x^{-1/2} \sum_{r=0}^{\infty} \frac{(-1)^r}{(2r)!} x^{2r}$$

It follows that the general solution in series about $x = 0$ is

$$y = Ax^{1/2} \sum_{r=0}^{\infty} \frac{(-1)^r}{(2r+1)!} x^{2r} + Bx^{-1/2} \sum_{r=0}^{\infty} \frac{(-1)^r}{(2r)!} x^{2r}$$

Since the differential equation has singular points only at $x = 0$, the series solutions are convergent for $|x| > 0$. The series resulting from $x = \frac{1}{2}$ is convergent with $x = 0$; the series resulting from $x = -\frac{1}{2}$ is not.

Example 6.9 Find the general solution in series about $x = 0$ of

$$x^2 y'' + xy' + (x^2 - 1)y = 0$$

We note that this differential equation is similar to that studied in the previous example. The resulting equations for m and the coefficients are

coefficient of x^m $(m^2 - 1)A_0 = 0$
coefficient of x^{m+1} $[(m+1)^2 - 1]A_1 = 0$
coefficient of x^{m+2} $[(m+2)^2 - 1]A_2 + A_0 = 0$
coefficient of x^{m+n} $[(m+n)^2 - 1]A_n + A_{n-2} = 0$

Hence $m = \pm 1$, and the odd indexed coefficients are zero.

(a) $m = 1$

$$A_2 = -\frac{A_0}{4 \cdot 2} = -\frac{A_0}{2^2 \, 2!} \qquad A_n = -\frac{A_{n-2}}{(n+2)n}$$

$$A_4 = -\frac{A_2}{6 \cdot 4} = \frac{A_0}{2^2 \cdot 3 \cdot 2 \cdot 2^2 \, 2!} = \frac{A_0}{2^4 \cdot 2 \cdot 3!}$$

$$A_6 = -\frac{A_4}{8 \cdot 6} = -\frac{A_0}{2^2 \cdot 4 \cdot 3 \cdot 2^4 \cdot 2 \cdot 3!} = -\frac{A_0}{2^6 \cdot 3! 4!}$$

and generally

$$A_n = \frac{(-1)^{n/2} A_0}{2^n (\frac{1}{2}n)!(\frac{1}{2}n + 1)!} \qquad n \text{ even}$$

Thus a solution is

$$y_1 = x\left[1 - \frac{x^2}{2^2 \, 2!} + \frac{x^4}{2^4 \, 2!3!} - \cdots\right] = x \sum_{r=0}^{\infty} \frac{(-1)^r x^{2r}}{2^{2r} r!(r+1)!}$$

(b) $m = -1$ The odd indexed coefficients are again zero, but we now note that

$$A_2 = -\frac{A_0}{(m+2)^2 - 1} \quad \text{with } m = -1$$

which is indeterminate since the denominator is zero.

Hence the method fails and we must look for a different method leading to another series so that we can obtain the general solution. It must be stressed, however, that the series we have obtained with $m = 1$ is a solution although it is not the general solution.

The roots m_1, m_2 ($m_1 \geqslant m_2$) of the indicial equation fit into one of three possible cases:

(a) the roots are unequal and do not differ by an integer, that is

$$m_1 > m_2 \qquad (m_1 - m_2) \text{ non-integral}$$

(b) the roots are unequal and differ by an integer, that is

$$m_1 > m_2 \qquad (m_1 - m_2) \text{ integral}$$

(c) the roots are equal, that is

$$m_1 = m_2$$

We consider these cases in turn.

(a) $m_1 > m_2$; $(m_1 - m_2)$ non-integral
   Two independent series solutions

$$y_1 = x^{m_1} \sum A_r x^r \quad \text{and} \quad y_2 = x^{m_2} \sum A_r x^r$$

can always be found, as in Example 6.7, so that the general series solution is

$$y = Ay_1 + By_2$$

(b) $m_1 > m_2$; $(m_1 - m_2)$ integral

In some cases it is possible to find two independent series solutions as in (a) above. In the other cases it is only possible to find one independent series solution corresponding to the larger root m_1 of the indicial equation. That is, a series solution is

$$y_1 = x^{m_1} \sum A_r x^r \tag{6.6}$$

Series Solution   149

but the second independent series solution can be shown to have the form

$$y_2 = y_1 \ln x + x^{m_2} \sum_{r=0}^{\infty} B_r x^r \qquad (6.7)$$

The coefficients B_r can be determined by substitution of y_2, y_2' and y_2'' into the differential equation and equating coefficients of powers of x to zero.

Now

$$y_2' = \frac{1}{x} y_1 + y_1' \ln x + \sum_{r=0}^{\infty} B_r(m_2 + r) x^{m_2 + r - 1}$$

$$y_2'' = -\frac{1}{x^2} y_1 + \frac{2}{x} y_1' + y_1'' \ln x + \sum_{r=0}^{\infty} B_r(m_2 + r)(m_2 + r - 1) x^{m_2 + r - 2}$$

and it is to be noted that when these series are substituted into the differential equation, the terms involving $\ln x$ disappear, since y_1 is a solution of the differential equation. Hence

$$[P_0(x) y_1'' + P_1(x) y_1' + P_2(x) y_1] \ln x = 0$$

It follows that it is not necessary to find y_1'' when evaluating the coefficients B_r. Note also that the terms resulting from the series on the right-hand side of (6.7) can be obtained from those already found for the series $x^m \sum A_r x^r$ by replacing m by m_2 and A_r by B_r. This process is illustrated by completing the solution to Example 6.9.

Example 6.9 (continued)   It has already been shown that $m_1 = 1$, $m_2 = -1$,

$$y_1 = x \sum_{r=0}^{\infty} \frac{(-1)^r x^{2r}}{2^{2r} r! (r+1)!} = x \left[1 - \frac{x^2}{2^2 \, 2!} + \frac{x^4}{2^4 \, 2! 3!} - \cdots \right]$$

and that a second independent solution of the form $x^m \sum A_r x^r$ does not exist.

A second independent solution is therefore of the form

$$y_2 = y_1 \ln x + x^{-1} \sum_{r=0}^{\infty} B_r x^r$$

Differentiating y_1, we obtain

$$y_1' = \sum_{r=0}^{\infty} \frac{(2r+1)(-1)^r x^{2r}}{2^{2r} r! (r+1)!} = 1 - \frac{3x^2}{2^2 \, 2!} + \frac{5x^4}{2^4 \, 2! 3!} - \cdots$$

150 Ordinary Differential Equations

Hence on substituting y_2'', y_2' and y_2 into the differential equation $x^2 y_2'' + xy_2' + (x^2 - 1)y_2 = 0$, the resulting coefficients of powers of x are as given in Table 6.8.

Note. The first row in the table is obtained by taking the set of equations for m and the coefficients A_r obtained previously (p. 147) and putting $m = m_2 = -1$ and $A_r = B_r$. For this particular example, the general term in the table depends upon whether the power of x is even or odd, and the sum of the second and fourth rows in the table is identically zero.

On equating coefficients of powers of x to zero, we obtain

$$B_1 = 0$$

$$B_0 = -2$$

$$3.1 B_3 + B_1 = 0$$

$$4.2 B_4 + B_2 = \frac{2.3}{2^2 2!}$$

$$(2n + 1)(2n - 1) B_{2n+1} + B_{2n-1} = 0$$

$$(2n + 2) 2n B_{2n+2} + B_{2n} = -\frac{(-1)^n 2(2n + 1)}{2^{2n} n!(n + 1)!}$$

It follows that $0 = B_1 = B_3 = B_5 = \cdots = B_{2n+1} = \cdots$, and that B_2 can take any value. Hence

$$B_0 = -2$$

$$B_4 = -\frac{B_2}{2^2 2!} + \frac{1}{2^3 2!} \left(\frac{3}{1.2} \right)$$

$$B_6 = -\frac{B_4}{2^2 3.2} - \frac{2.5}{2^4 3! 2! 2^2 3.2}$$

$$= \frac{B_2}{2^4 3! 2!} - \frac{1}{2^5 3! 2!} \left(\frac{3}{1.2} + \frac{5}{2.3} \right)$$

and similarly

$$B_8 = -\frac{B_2}{2^6 4! 3!} + \frac{1}{2^7 4! 3!} \left(\frac{3}{1.2} + \frac{5}{2.3} + \frac{7}{3.4} \right)$$

Table 6.8

Term	x^{-1}	Const	x	x^2	x^3	x^{2n}	x^{2n+1}
$x^{-1}\sum B_r x^r$	0	$-B_1$	B_0	$3.1B_3+B_1$	$4.2B_4+B_2$	$(2n+1)(2n-1)B_{2n+1}+B_{2n-1}$	$(2n+2)2nB_{2n+2}+B_{2n}$
$x^2 y_2'' \begin{cases} -y_1 \\ 2xy_1' \end{cases}$			-1	0	$\dfrac{1}{2^2 2!}$	0	$-\dfrac{(-1)^n}{2^{2n} n!(n+1)!}$
			2	0	$-\dfrac{2\cdot 3}{2^2 2!}$	0	$\dfrac{(-1)^n 2(2n+1)}{2^{2n} n!(n+1)!}$
$xy_2'\{y_1$			1	0	$-\dfrac{1}{2^2 2!}$	0	$\dfrac{(-1)^n}{2^{2n} n!(n+1)!}$

Hence

$$x^{-1}\sum_{r=0}^{\infty}B_r x^r = \frac{1}{x}\left[-2 + \frac{1}{2^3 2!}\left(\frac{3}{1.2}\right)x^4\right.$$

$$-\frac{1}{2^5 3!2!}\left(\frac{3}{1.2}+\frac{5}{2.3}\right)x^6 + \cdots$$

$$+ B_2\left(x^2 - \frac{1}{2^2 2!}x^4 + \frac{1}{2^4 3!2!}x^6 - \cdots\right)\right]$$

$$= -\frac{2}{x}\left[1 - \frac{1}{2^4 2!}\frac{3}{1.2}x^4\right.$$

$$\left. + \frac{1}{2^6 3!2!}\left(\frac{3}{1.2}+\frac{5}{2.3}\right)x^6 - \cdots\right] + B_2 y_1$$

The second independent solution is therefore

$$y_2 = y_1(B_2 + \ln x)$$

$$-\frac{2}{x}\left[1 - \frac{1}{2^4 2!}\left(\frac{3}{1.2}\right)x^4 + \frac{1}{2^6 3!2!}\left(\frac{3}{1.2}+\frac{5}{2.3}\right)x^6 - \cdots\right]$$

Since there is a singular point at $x = 0$ only, the series solutions y_1 and y_2 are convergent for $|x| > 0$. The series y_1 is obviously convergent for $x = 0$, and since y_2 involves $\ln x$ it is not convergent for $x = 0$. The general solution is given by

$$y = Ay_1 + By_2$$

$$= (A + BB_2 + B \ln x)y_1 - \frac{2B}{x}\left[1 - \frac{1}{2^4 2!}\left(\frac{3}{1.2}\right)x^4 + \cdots\right]$$

It is to be noted that the two arbitrary constants are $A + BB_2$ and B, so that in effect the constant B_2 can be taken as zero, or indeed any convenient number.

This applies similarly to other examples. *Hence in obtaining a second independent series solution of the form (6.7), the coefficient $B_{(m_1-m_2)}$ can be given any conveniently chosen numerical value, and will often be taken as zero.*

(c) If the roots are equal, then obviously only one solution of the form $x^m \Sigma A_r x^r$ can be found, and the other solution is obtained in a similar manner to that for roots differing by an integer. That is, another solution is given by (6.7), where y_1 is the series solution of the form (6.6), $m_1 = m_2$, and B_0 can take any value.

Example 6.10 Find the general solution in series about $x = 0$ of
$$xy'' + y' + xy = 0$$
$x = 0$ is a regular singular point, so that a solution of the form
$$y = x^m \sum_{r=0}^{\infty} A_r x^r \qquad A_0 = 1$$
exists.

In the usual way the equations for m and A are

Coefficient of x^{m-1} $m^2 A_0 = 0$ (double root at $m = 0$)

Coefficient of x^m $(m+1)^2 A_1 = 0$

Coefficient of x^{m+1} $(m+2)^2 A_2 + A_0 = 0$

Coefficient of x^{m+2} $(m+3)^2 A_3 + A_1 = 0$

Coefficient of x^{m+n} $(m+n+1)^2 A_{n+1} + A_{n-1} = 0$

Hence a solution with $m = 0$ is
$$y_1 = 1 - \frac{x^2}{2^2} + \frac{x^4}{2^4 (2!)^2} - \frac{x^6}{2^6 (3!)^2} + \cdots$$

Another independent solution is given by
$$y_2 = y_1 \ln x + x^0 \sum_{r=0}^{\infty} B_r x^r$$

The coefficients of powers of x on substituting y_2'', y_2' and y_2 into the differential equation are as shown in Table 6.9.

The first row is obtained from the equations for coefficients of powers of x with m put equal to zero. General columns are needed for both odd and even powers of x, since the series for y_1 involves only even powers of x. Hence
$$0 = B_1 = B_3 = B_5 = \cdots = B_{2n+1} = \cdots$$
and as explained earlier, since B_0 can take any value, we take it to be zero.

Table 6.9

Term	x^{-1}	Const.	x	x^{2n-1}	x^{2n-1}	x^{2n}
$\sum B_r x^r$	0	B_1	$2^2 B_2 + B_0$	$3^2 B_3 + B_1$	$(2n)^2 B_{2n} + B_{2n-2}$	$(2n+1)^2 B_{2n+1} + B_{2n-1}$
$xy_2'' \begin{cases} -\dfrac{1}{x} y_1 \\ 2y_1' \end{cases}$	-1	0	$\dfrac{1}{2^2}$	0	$-\dfrac{(-1)^n}{2^{2n}(n!)^2}$	0
	0	0	$-\dfrac{2 \cdot 2}{2^2}$	0	$\dfrac{(-1)^n 2 \cdot 2n}{2^{2n}(n!)^2}$	0
$y' \left\{ \dfrac{1}{x} y_1 \right.$	1	0	$-\dfrac{1}{2^2}$	0	$\dfrac{(-1)^n}{2^{2n}(n!)^2}$	0

Hence

$$2^2 B_2 = \frac{2 \cdot 2}{2^2} \qquad B_2 = \frac{1}{2^2}.$$

$$4^2 B_4 + B_2 = -\frac{2 \cdot 4}{2^4 (2!)^2} \qquad B_4 = -\frac{1}{2^4 (2!)^2}(1 + \tfrac{1}{2})$$

$$6^2 B_6 + B_4 = \frac{2 \cdot 6}{2^6 (3!)^2} \qquad B_6 = \frac{1}{2^6 (3!)^2}(1 + \tfrac{1}{2} + \tfrac{1}{3})$$

Hence the second independent solution is

$$y_2 = \ln x \left[1 - \frac{x^2}{2^2} + \frac{x^4}{2^4 (2!)^2} - \frac{x^6}{2^6 (3!)^2} + \cdots \right]$$

$$+ \left[\frac{x^2}{2^2} - \frac{x^4}{2^4 (2!)^2}(1 + \tfrac{1}{2}) + \frac{x^6}{2^6 (3!)^2}(1 + \tfrac{1}{2} + \tfrac{1}{3}) - \cdots \right]$$

The general solution is

$$y = (A + B \ln x) \sum_{r=0}^{\infty} \frac{(-1)^r x^{2r}}{2^{2r}(r!)^2} + B \sum_{r=1}^{\infty} \frac{(-1)^{r-1} x^{2r}}{2^{2r}(r!)^2} h_r$$

where

$$h_r = \left(1 + \tfrac{1}{2} + \tfrac{1}{3} + \cdots + \tfrac{1}{r}\right)$$

The series is convergent for $|x| > 0$.

Summary of solution in series about the regular singular point, $x = 0$, of the differential equation

$$P_0(x) \frac{d^2 y}{dx^2} + P_1(x) \frac{dy}{dx} + P_2(x) y = 0 \qquad P_0(0) = 0$$

Let roots of indicial equation be m_1 and m_2.

(a) $m_1 > m_2$, $(m_1 - m_2)$ non-integral

Two independent series solutions exist of the form

$$y_1 = x^{m_1} \sum_{r=0}^{\infty} A_r x^r \quad \text{and} \quad y_2 = x^{m_2} \sum_{r=0}^{\infty} A_r x^r \quad \text{with } A_0 = 1$$

(b) $m_1 > m_2$, $(m_1 - m_2)$ integral

One solution exists of the form

$$y_1 = x^{m_1} \sum A_r x^r \qquad A_0 = 1$$

A second independent solution of the form $y_2 = x^{m_2} \sum A_r x^r$ may exist; if it does not, then the form for y_2 is

$$y_2 = y_1 \ln x + x^{m_2} \sum_{r=0}^{\infty} B_r x^r$$

where $B_{(m_1 - m_2)} = 0$ (or any other convenient number).

(c) $m_1 = m_2$

The forms of the two independent solutions are

$$y_1 = x^{m_1} \sum_{r=0}^{\infty} A_r x^r \qquad A_0 = 1$$

and

$$y_2 = y_1 \ln x + x^{m_2} \sum_{r=0}^{\infty} B_r x^r$$

with $m_1 = m_2$ and $B_0 = 0$ (or any other convenient number).

In all cases, the general solution is

$$y = Ay_1 + By_2$$

We now consider, briefly, the solution of the differential equation

$$P_0(x) \frac{d^2 y}{dx^2} + P_1(x) \frac{dy}{dx} + P_2(x) y = Q(x)$$

Since the equation is linear, the solution, as for those differential equations considered in Section 4.3, is the general solution of the homogeneous equation plus a particular integral. The particular integral can sometimes be obtained by inspection and sometimes by setting $y = x^m \sum D_r x^r$ and equating powers of x on the left-hand side to those on the right-hand side of the equation. The particular integral for each term on the right-hand side of the equation is taken in turn.

Example 6.11 Find a particular integral of

$$x^2 y'' + xy' + (x^2 - 1)y = x^2$$

The solution of the homogeneous equation has been found in Example 6.9. The coefficients of powers of x on the left-hand side of the equation, with $y = x^m \sum A_r x^r$, have been obtained in Example 6.9, so that on equating powers of x on both sides of the equation, after replacing A_r by D_r, we obtain

$$(m^2 - 1)D_0 x^m \equiv x^2$$
$$[(m + 1)^2 - 1]D_1 x^{m+1} \equiv 0$$
$$\{[(m + 2)^2 - 1]D_2 + D_0\}x^{m+2} \equiv 0$$
$$\{[(m + n)^2 - 1]D_n + D_{n-2}\}x^{m+n} \equiv 0$$

From the first identity, $m = 2$ and $D_0 = \frac{1}{3}$.
With these values the other identities give

$$0 = D_1 = D_3 = D_5 = \cdots$$

and

$$5 \cdot 3D_2 + D_0 = 0 \qquad D_2 = -\frac{1}{5 \cdot 3^2} = -\frac{2^2}{5(3!)^2}$$

$$7 \cdot 5D_4 + D_2 = 0 \qquad D_4 = \frac{1}{7 \cdot 5^2 3^2} = +\frac{2^4(2!)^2}{7(5!)^2}$$

$$(n + 3)(n + 1)D_n + D_{n-2} = 0 \qquad D_n = \frac{(-1)^{n/2}}{(n + 3)(n + 1)^2 \cdots 3^2}$$

The even-values coefficients are therefore given by

$$D_{2n} = \frac{(-1)^n 2^{2n}(n!)^2}{(2n + 3)[(2n + 1)!]^2}$$

The particular integral is

$$y = x^2 \sum_{n=0}^{\infty} \frac{(-1)^n 2^{2n}(n!)^2}{(2n + 3)[(2n + 1)!]^2} x^{2n}$$

and the series is convergent for all x.

Problem 6.6 Show that $x = 0$ is a regular singular point of the following differential equations, and determine the range of x for which the solutions in series are convergent
(a) $x^2(x^2 - 1)y'' + xy' - y = 0$
(b) $x(2x + 1)y'' - y' + xy = 0$
(c) $x(x - 1)(2x^2 + x + 1)y'' + y' + y = 0$

Problem 6.7 Find the general solution in series about $x = 0$, or $t = 0$, stating the range of convergence, for the following differential equations:

(a) $3xy'' + y' - (5 + 3x)y = 0$

(b) $x(1 - x)y'' + 2(1 - 2x)y' - 2y = 0$

(c) $tx'' - 3x' + tx = 0$

(d) $t^2 x'' - tx' + (t^2 + 1)x = 0$

6.5 Special differential equations

The following differential equations have important applications in mathematical physics: for example, in heat flow, vibration of membranes, fluid flow and electrostatics.

(a) $x^2 \dfrac{d^2 y}{dx^2} + x \dfrac{dy}{dx} + (x^2 - k^2)y = 0$

(b) $(1 - x^2) \dfrac{d^2 y}{dx^2} - 2x \dfrac{dy}{dx} + p(p + 1)y = 0$

The first equation is known as *Bessel's equation* and the second as *Legendre's equation*.

6.5.1 Bessel's equation and Bessel functions

For real values of k the differential equation

$$x^2 \frac{d^2 y}{dx^2} + x \frac{dy}{dx} + (x^2 - k^2)y = 0$$

is known as Bessel's equation of *order k*. Let us consider Bessel's equation of order zero. This is the differential equation that was studied in Example 6.10.

The solution in series about $x = 0$ was found to be

$$y = (A + B \ln x) \sum_{r=0}^{\infty} \frac{(-1)^r}{(r!)^2} \left(\frac{x}{2}\right)^{2r} + B \sum_{r=1}^{\infty} \frac{(-1)^{r-1} h_r}{(r!)^2} \left(\frac{x}{2}\right)^{2r}$$

with
$$h_r = 1 + \tfrac{1}{2} + \tfrac{1}{3} + \cdots + \frac{1}{r}$$

The solution can be rewritten as
$$y = a \sum_{r=0}^{\infty} \frac{(-1)^r}{(r!)^2}\left(\frac{x}{2}\right)^{2r} + b\,\frac{2}{\pi}\left\{\left(\ln\frac{x}{2}+\gamma\right)\sum_{r=0}^{\infty}\frac{(-1)^r}{(r!)^2}\left(\frac{x}{2}\right)^{2r} \right.$$
$$\left. + \sum_{r=1}^{\infty} \frac{(-1)^{r-1} h_r}{(r!)^2}\left(\frac{x}{2}\right)^{2r}\right\}$$

where γ is Euler's constant = 0·5772 (to four decimal places) and where
$$B = \frac{2b}{\pi} \quad\text{and}\quad A = a + \frac{2b}{\pi}(\gamma - \ln 2)$$

The general solution to Bessel's equation of order zero can therefore be written as
$$y = aJ_0(x) + bY_0(x)$$

where
$$J_0(x) = \sum_{r=0}^{\infty} \frac{(-1)^r}{(r!)^2}\left(\frac{x}{2}\right)^{2r}$$

and
$$Y_0(x) = \frac{2}{\pi}\left[J_0(x)\left(\gamma + \ln\frac{x}{2}\right) + \sum_{r=1}^{\infty} \frac{(-1)^{r-1} h_r}{(r!)^2}\left(\frac{x}{2}\right)^{2r}\right]$$

$J_0(x)$ is Bessel's function of the first kind of order zero.
$Y_0(x)$ is Bessel's function of the second kind of order zero.
The series expression for $J_0(x)$ is convergent for all x, while the series expression for $Y_0(x)$ is convergent for $x \neq 0$. $J_0(x)$ and $Y_0(x)$ are solutions of Bessel's equation with $k = 0$; this is analogous to the way in which solutions of the differential equation
$$\frac{d^2 y}{dx^2} + y = 0$$
are
$$\sin x = x - \frac{x^3}{3!} + \frac{x^5}{5!} - \cdots$$

160   Ordinary Differential Equations

and

$$\cos x = 1 - \frac{x^2}{2!} + \frac{x^4}{4!} - \cdots$$

The circular functions are well tabulated, and their differential, integral and identity properties are well known. In a similar way the Bessel functions are well tabulated and their differential, integral and identity properties are well documented. We next consider the Bessel function of order one which was studied in Example 6.9. The series solution obtained can be rewritten so that the solution is

$$y(x) = aJ_1(x) + bY_1(x)$$

where
$J_1(x)$ is Bessel's function of the first kind of order one
$Y_1(x)$ is Bessel's function of the second kind of order one and where

$$J_1(x) = \left(\frac{x}{2}\right) \sum_{r=0}^{\infty} \frac{(-1)^r}{r!(r+1)!} \left(\frac{x}{2}\right)^{2r}$$

and

$$Y_1(x) = \frac{2}{\pi}\left(\gamma + \ln\frac{x}{2}\right)J_1(x) - \frac{2}{\pi x}$$
$$- \frac{1}{\pi}\left(\frac{x}{2}\right) \sum_{r=0}^{\infty} \frac{(-1)^r}{r!(r+1)!} \left(\frac{x}{2}\right)^{2r} \{h_r + h_{r+1}\}$$

where $h_0 = 0$.

We now consider a Bessel equation of non-integral order, namely of order $\frac{1}{2}$, the solution of which was obtained in Example 6.8 and was

$$y(x) = Ax^{1/2} \sum_{r=0}^{\infty} \frac{(-1)^r}{(2r+1)!} x^{2r} + Bx^{-1/2} \sum_{r=0}^{\infty} \frac{(-1)^r}{r!} x^{2r}$$

This can be rewritten to give

$$y(x) = aJ_{1/2}(x) + bJ_{-1/2}(x)$$

where $J_{1/2}(x)$ and $J_{-1/2}(x)$ are Bessel functions of the first kind of order $\frac{1}{2}$ and $-\frac{1}{2}$ respectively, and where

$$J_{1/2}(x) = \left(\frac{x}{2}\right)^{1/2} \sum_{r=0}^{\infty} \frac{(-1)^r}{r!\,\Gamma(r+\frac{3}{2})} \left(\frac{x}{2}\right)^{2r}$$

and
$$J_{-1/2}(x) = \left(\frac{x}{2}\right)^{-1/2} \sum_{r=0}^{\infty} \frac{(-1)^r}{r!\,\Gamma(r+\tfrac{1}{2})} \left(\frac{x}{2}\right)^{2r}$$
where $\Gamma(n)$ is the gamma function, which is such that
$$\Gamma(n+1) = n\Gamma(n) \text{ and } \Gamma(\tfrac{1}{2}) = \sqrt{\pi}$$
(Barnett and Cronin (1975), Section 4.4).

6.5.2. Legendre's equation and Legendre's polynomials

The series solution for Legendre's equation with $p = 2$ has already been found in Example 6.3. It is noted that one of the two independent solutions is a terminated series, namely
$$y(x) = 3x^2 - 1$$
If p is any positive integer n, then it is found that one of the two independent series solutions is always a polynomial. These polynomials multiplied by appropriate constants are called Legendre polynomials and denoted by $P_n(x)$.

A list of some of the Legendre polynomials is
$$P_0(x) = 1$$
$$P_1(x) = x$$
$$P_2(x) = \tfrac{1}{2}(3x^2 - 1)$$
$$P_3(x) = \tfrac{1}{2}(5x^3 - 3x)$$
$$P_4(x) = \frac{1}{2\cdot 4}(5\cdot 7x^4 - 2\cdot 3\cdot 5x^2 + 1\cdot 3)$$

and in general
$$P_n(x) = \frac{(2n-1)(2n-3)\cdots 1}{n!}\left[x^n - \frac{n(n-1)}{2(2n-1)}x^{n-2} + \frac{n(n-1)(n-2)(n-3)}{2\cdot 4(2n-1)(2n-3)}x^{n-4} - \cdots\right]$$

The Legendre polynomials are chosen such that as well as being solutions to Legendre's equation they have the property that $P_n(1) = 1$. When p is non-integral, neither of the two series is terminated. An example of this has been evaluated in Problem 6.4, where $p = \tfrac{1}{2}$.

Problem 6.8 Show that a solution of the Bessel equation of order k is given by

$$J_k(x) = \left(\frac{x}{2}\right)^k \sum_{r=0}^{\infty} \frac{(-1)^r}{r!\,\Gamma(k+r+1)} \left(\frac{x}{2}\right)^{2r}$$

Problem 6.9 Show that the general solution of the *modified* Bessel equation of order one

$$x^2 y'' + xy' - (x^2 + 1)y = 0$$

is

$$y = AI_1(x) + BK_1(x)$$

where the modified Bessel function of the first kind of order one, $I_1(x)$, is

$$I_1(x) = \left(\frac{x}{2}\right) \sum_{r=0}^{\infty} \frac{1}{r!(r+1)!} \left(\frac{x}{2}\right)^{2r}$$

and the modified Bessel function of the second kind of order one, $K_1(x)$, is

$$K_1(x) = \left\{\gamma + \ln\frac{x}{2}\right\} I_1(x) + \tfrac{1}{2}\left(\frac{2}{x}\right)$$

$$- \tfrac{1}{2}\left(\frac{x}{2}\right) \sum_{r=0}^{\infty} \frac{1}{r!(r+1)!} \left(\frac{x}{2}\right)^{2r} \{h_r + h_{r+1}\}$$

Problem 6.10 Using the series expansion of $J_k(x)$, prove the following:

(a) $\dfrac{d}{dx} J_0(x) = -J_1(x)$

(b) $\dfrac{d}{dx} [x^k J_k(x)] = x^k J_{k-1}(x)$

(c) $\dfrac{d}{dx} [x^{-k} J_k(x)] = -x^{-k} J_{k+1}(x)$

Problem 6.11 Using the results obtained in Problem 6.10, show that

$$2k J_k(x) = x[J_{k-1}(x) + J_{k+1}(x)]$$

Problem 6.12 Prove the following:

(a) $\int \dfrac{J_4(x)}{x^3} dx = \left(\dfrac{1}{x^3} - \dfrac{8}{x^5}\right) J_1(x) + \dfrac{4}{x^2} J_0(x) + \text{constant}$

(Hint: Use results of Problems 6.10 and 6.11)

(b) $\int xJ_1(x) \cos x\, dx = xJ_1(x) \sin x - \int x \sin x\, J_0(x)\, dx$

(c) $\int J_0(x) \cos x\, dx = xJ_0(x) \cos x + xJ_1(x) \sin x + \text{constant}$

Problem 6.13 The Legendre polynomials satisfy the following relations:

(a) $(n+1)P_{n+1}(x) = (2n+1)xP_n(x) - nP_{n-1}(x)$

(b) $(x^2 - 1)P'_n(x) = nxP_n(x) - nP_{n-1}(x)$

(c) $P_n(x) = \dfrac{1}{2^n n!} \dfrac{d^n}{dx^n} (x^2 - 1)^n$

Prove that these hold true when $n = 3$ and 4.

Problem 6.14 The Legendre polynomials are orthogonal polynomials satisfying the following

$$\int_{-1}^{1} P_n(x) P_m(x)\, dx = 0 \qquad n \neq m$$

$$= \dfrac{2}{2n+1} \qquad m = n$$

Verify this when $n = 2$ and 3 and $m = 1$ and 2.

Miscellaneous problems

Problem 6.15 Find solutions in series about $x = 0$, or $t = 0$, of the following differential equations:

(a) $y/y' - yy' = 2x$, with $y = 1$ when $x = 0$;

  obtain series up to and including the term x^3

(b) $yy' + y^2 = x^2$, with $y \neq 0$ when $x = 0$;

  obtain series up to and including the term x^3

(c) $y'' + xy = 0$; state the radius of convergence
(d) $x^2 y'' + 4xy' + (x^2 + 2)y = 0$
(e) $tx'' + (1 - 2t)x' + (t - 1)x = 0$
(f) $(t^2 - 1)t^2 x'' - (t^2 + 1)tx' + (t^2 + 1)x = 0$

Problem 6.16 Find the solution in series about $x = 1$ of the differential equation
$$2x(x-1)^2 y'' + (3x - 4)(x - 1)y' + y = 0$$

Problem 6.17 Find the solution in series valid for large x of the differential equation
$$x^4 y'' - 4x^3 y' + (6x^2 - 1)y = 0$$

CHAPTER SEVEN

Laplace Transforms

7.1 Introduction

In this chapter we shall obtain the solution of linear ordinary differential equations or systems of such equations by means of the Laplace transform.

The use of the Laplace transform can be illustrated by considering the equation

$$\frac{dx}{dt} = ae^{-at} \qquad \text{with } x = 1 \text{ when } t = 0$$

Multiply each side of the equation by e^{-st}, where s is a parameter, and integrate with respect to t from zero to infinity. Then provided that the integrals exist,

$$\int_0^\infty e^{-st} \frac{dx}{dt} dt = a \int_0^\infty e^{-st} e^{-at} dt = a \int_0^\infty e^{-t(s+a)} dt$$

Hence, if the integral on the left-hand side of the equation is integrated by parts,

$$[x(t)e^{-st}]_0^\infty + s \int_0^\infty x(t)e^{-st} = a \left[-\frac{e^{-t(s+a)}}{s+a} \right]_0^\infty$$

The dependent variable x is written as $x(t)$ to stress that it is a function of the independent variable t. The parameter s is assumed to be such that the contribution of both the upper limits of the square brackets are zero, so that if we define the *Laplace transform*, $\bar{x}(s)$ of $x(t)$ as

$$\bar{x}(s) = \int_0^\infty e^{-st} x(t)\, dt$$

then the above equation becomes, on using the initial condition,

$$s\bar{x}(s) - 1 = a/(s+a)$$

The transform is written as $\bar{x}(s)$ to stress that it is a function of the parameter s.

The differential equation has thus been transformed into an algebraic equation which can be solved to give

$$\bar{x}(s) = \frac{1}{s} + \frac{a}{s(s+a)} = \frac{2}{s} - \frac{1}{(s+a)}$$

The solution of the differential equation, $x(t)$, is the inverse Laplace transform of $\bar{x}(s)$. The transform of e^{-at} has already been shown to be $1/(s+a)$, so the inverse Laplace transform of $1/(s+a)$ is e^{-at}. It follows, on putting $a = 0$, that the inverse of $1/s$ is unity and hence the solution of the equation is

$$x(t) = 2 - e^{-at}$$

Tables of Laplace transforms and its inverse have been constructed, so that with these and a few basic rules, the Laplace transform or the inverse Laplace transform can be readily obtained for most functions that arise in practice.

The steps in the solution of an ordinary linear differential equation or system of linear differential equations can be summarised as follows:

(a) The differential equation (or system of differential equations) is transformed into an algebraic equation (or system of algebraic equations) by means of the Laplace transform.
(b) The equation (or equations) is solved for the transform of the dependent variable (or variables).
(c) Using tables of inverse Laplace transforms, the solution of the differential equation (or system of differential equations) is obtained.

7.2 The Laplace transform

The Laplace transform of the function $x(t)$, $t \geq 0$, is denoted either by $\bar{x}(s)$ or $\mathcal{L}[x(t)]$ and defined by

$$\bar{x}(s) = \mathcal{L}[x(t)] = \int_0^\infty e^{-st} x(t)\, dt \qquad (7.1)$$

which assumes that the integral exists.

The Laplace transform of a few elementary functions are now derived using the definition.

(a)
$$\mathcal{L}[1] = \int_0^\infty e^{-st}\, dt = \left[-\frac{1}{s} e^{-st}\right]_\infty^0 = \frac{1}{s} \qquad (7.2)$$

assuming that $e^{-st} \to 0$ as $t \to \infty$. That is, the parameter s must be such that $s > 0$.

(b) If n is a positive integer,

$$\mathcal{L}[t^n] = \int_0^\infty t^n e^{-st}\, dt = \frac{n}{s} \int_0^\infty t^{n-1} e^{-st}\, dt$$

assuming again that $s > 0$. Hence

$$\mathcal{L}[t^n] = \frac{n(n-1)}{s^2} \int_0^\infty t^{n-2} e^{-st}\, dt = \cdots = \frac{n!}{s^n} \int_0^\infty 1 \cdot e^{-st}\, dt = \frac{n!}{s^{n+1}} \qquad (7.3)$$

(c)
$$\mathcal{L}[e^{at}] = \int_0^\infty e^{at} e^{-st}\, dt = \int_0^\infty e^{-(s-a)t}\, dt = \frac{1}{s-a} \qquad (7.4)$$

assuming that $s > a$.

Henceforth it will be assumed that the value of s has been chosen such that the integrals exist.

(d)
$$\mathcal{L}[\sin t] = \int_0^\infty \sin t\, e^{-st}\, dt$$

$$= \left[\frac{e^{-st}(-s \sin t - \cos t)}{s^2 + 1}\right]_0^\infty = \frac{1}{s^2 + 1} \qquad (7.5)$$

(See Barnett and Cronin (1975), Section 3.4.4.)

(e) $$\mathcal{L}[\cos t] = \int_0^\infty \cos t \, e^{-st} \, dt$$
$$= \left[\frac{e^{-st}(-s\cos t + \sin t)}{s^2 + 1} \right]_0^\infty = \frac{s}{s^2 + 1} \qquad (7.6)$$

(See Barnett and Cronin (1975), Section 3.4.5.)

The Laplace transform of any function $x(t)$ can be obtained by direct integration, but their evaluation can be facilitated by use of tables of Laplace transforms and by use of certain basic rules.

There are many large tables of Laplace transforms, and a short useful table is given in Barnett and Cronin (1975), Section 6.5, part of which is reproduced in an appendix to this chapter. The basic rules useful in the extension of any table are listed and proved in the next section.

Problem 7.1 Verify the following results by direct integration:

(a) $\mathcal{L}[te^{-at}] = 1/(s+a)^2$

(b) $\mathcal{L}[e^{-2t} \sin 3t] = 3/(s^2 + 4s + 13)$

(c) $\mathcal{L}[f(t)] = (1 - e^{-s})/s^2$,

where
$$f(t) = \begin{cases} t & 0 < t < 1 \\ 1 & t > 1 \end{cases}$$

7.3 Rules for Laplace transforms

In the following rules a, b are constants, $\bar{x}(s) = \mathcal{L}[x(t)]$ and $\bar{y}(s) = \mathcal{L}[y(t)]$.

7.3.1 Linearity

$$\mathcal{L}[ax(t) + by(t)] = a\bar{x}(s) + b\bar{y}(s) \qquad (7.7)$$

Proof
$$\mathcal{L}[ax(t) + by(t)] = \int_0^\infty [ax(t) + by(t)] e^{-st} \, dt$$
$$= a \int_0^\infty x(t) e^{-st} + b \int_0^\infty y(t) e^{-st} \, dt$$
$$= a\bar{x}(s) + b\bar{y}(s)$$

Example 7.1 Evaluate $\mathcal{L}[2 - e^{-at}]$

$$\mathcal{L}[2 - e^{-at}] = 2\mathcal{L}[1] - \mathcal{L}[e^{-at}] \quad \text{by (7.7)}$$

$$= \frac{2}{s} - \frac{1}{(s+a)} \quad \text{by (7.2) and (7.4)}$$

7.3.2 Change of scale

$$\mathcal{L}[x(at)] = \frac{1}{a} \bar{x}\left(\frac{s}{a}\right) \tag{7.8}$$

Proof

$$\mathcal{L}[x(at)] = \int_0^\infty x(at) e^{-st} \, dt$$

$$= \frac{1}{a} \int_0^\infty x(z) e^{-(s/a)z} \, dz \quad \text{(on putting } z = at\text{)}$$

$$= \frac{1}{a} \bar{x}\left(\frac{s}{a}\right)$$

since the integral is the definition (7.1) with s replaced by s/a.

Example 7.2 Evaluate $\mathcal{L}[\sin at]$.

(7.5) gives

$$\mathcal{L}[\sin t] = \bar{x}(s) = \frac{1}{(s^2 + 1)}$$

Hence, by (7.8),

$$\mathcal{L}[\sin at] = \frac{1}{a} \bar{x}\left(\frac{s}{a}\right) = \frac{1}{a} \frac{1}{(s^2/a^2) + 1} = \frac{a}{s^2 + a^2}$$

7.3.3 Differentiation

$$\mathcal{L}\left[\frac{dx}{dt}\right] = s\bar{x}(s) - x(0+) \tag{7.9}$$

where $x(0+) = \lim_{\epsilon \to 0} [x(\epsilon)] \quad \epsilon > 0$

Proof

$$\mathcal{L}\left[\frac{dx}{dt}\right] = \int_0^\infty e^{-st} \frac{dx}{dt} dt$$

$$= [x(t)e^{-st}]_0^\infty + s \int_0^\infty x e^{-st} dt \quad \text{(on integration by parts)}$$

$$= 0 - x(0+) + s\bar{x}(s)$$

assuming that $e^{-st}x(t) \to 0$ as $t \to \infty$, which must be satisfied if $\mathcal{L}[x(t)]$ is to exist.

7.3.4 Integration

$$\mathcal{L}\left[\int_0^t x(u) \, du\right] = \bar{x}(s)/s \qquad (7.10)$$

Proof. The proof depends upon the change of order of double integrals, and is given below for those readers familiar with this subject, but should be omitted by those who are not.

$$\mathcal{L}\left[\int_0^t x(u) \, du\right] = \int_0^\infty e^{-st} \left[\int_0^t x(u) \, du\right] dt$$

$$= \int_0^\infty x(u) \left[\int_u^\infty e^{-st} \, dt\right] du$$

(on changing the order of integration, which is permitted if $\bar{x}(s)$ exists)

$$= \frac{1}{s} \int_0^\infty x(u) e^{-su} \, du = \frac{\bar{x}(s)}{s}$$

Example 7.3 Given that $\mathcal{L}[\cos t] = s/(s^2 + 1)$, find $\mathcal{L}[\sin t]$.

We have
$$\sin t = \int_0^t \cos u \, du$$

so
$$\mathcal{L}[\sin t] = \mathcal{L}\left[\int_0^t \cos u \, du\right] = \frac{1}{s} \mathcal{L}[\cos t] = \frac{1}{(s^2 + 1)}$$

7.3.5 Multiplication by t

$$\mathcal{L}[tx(t)] = -\frac{d\bar{x}}{ds} \qquad (7.11)$$

Proof

$$\mathcal{L}[tx(t)] = \int_0^\infty tx(t)e^{-st}\,dt = -\int_0^\infty x(t)\frac{d}{ds}(e^{-st})\,dt$$

$$= -\frac{d}{ds}\int_0^\infty x(t)e^{-st}\,dt = -\frac{d}{ds}\bar{x}(s)$$

Example 7.4 Given that $\mathcal{L}[\sin t] = 1/(s^2 + 1)$, find $\mathcal{L}[t^2 \sin t]$.

To find the transform of the given function, it is necessary to use this rule twice.

$$\mathcal{L}[t^2 \sin t] = -\frac{d}{ds}\mathcal{L}[t\sin t] = -\frac{d}{ds}\left\{-\frac{d}{ds}\mathcal{L}[\sin t]\right\}$$

$$= \frac{d^2}{ds^2}\mathcal{L}[\sin t] = \frac{d^2}{ds^2}\left(\frac{1}{s^2+1}\right) = \frac{2(3s^2 - 1)}{(s^2+1)^3}$$

7.3.6 Division by t

$$\mathcal{L}\left[\frac{x(t)}{t}\right] = \int_s^\infty \bar{x}(u)\,du \qquad (7.12)$$

provided that $\lim_{t\to 0} x(t)/t$ exists.

Proof. The proof depends upon replacing e^{-st}/t by $\int_s^\infty e^{-ut}\,du$ and changing the order of integration of a double integral, and is left as a problem for those familiar with this subject.

Example 7.5 Given that $\mathcal{L}[\sin t] = 1/(s^2 + 1)$, find $\mathcal{L}[(\sin t)/t]$.

$$\mathcal{L}\left[\frac{\sin t}{t}\right] = \int_s^\infty \frac{du}{u^2 + 1}$$

$$= \frac{\pi}{2} - \tan^{-1} s = \cot^{-1} s$$

7.3.7 Higher derivatives

$$\mathcal{L}\left[\frac{d^n x}{dt^n}\right] = s^n x(s) - s^{n-1} x(0) - s^{n-2} x'(0) - \cdots - x^{(n-1)}(0) \quad (7.13)$$

where $x^r(0) = d^r x/dt^r$ evaluated at $t = \lim_{\epsilon \to 0} \epsilon, \epsilon > 0$.

This is a generalisation of (7.9), which is the above expression with $n = 1$.

Proof. Assume the expression true for n, then

$$\mathcal{L}\left[\frac{d^{n+1} x}{dt^{n+1}}\right] = \int_0^\infty \frac{d^{n+1} x}{dt^{n+1}} e^{-st}\, dt$$

$$= \left[\frac{d^n x}{dt^n} e^{-st}\right]_0^\infty + s \int_0^\infty \frac{d^n x}{dt^n} e^{-st}\, dt$$

$$= -x^n(0) + s[s^n \bar{x}(s) - s^{n-1} x(0) - \cdots - x^{(n-1)}(0)]$$
(by assumption)

$$= s^{n+1} \bar{x}(s) - s^n x(0) - \cdots - x^n(0)$$

Hence if the expression is true for n, then it is also true for $n + 1$. Since it has been shown to be true for $n = 1$, (7.9), it is therefore true for $n = 2$, hence for $n = 3$, and similarly for all positive integer values of n.

7.3.8 First shifting theorem

$$\mathcal{L}[e^{at} x(t)] = \bar{x}(s - a) \quad (7.14)$$

Proof

$$\mathcal{L}[e^{at} x(t)] = \int_0^\infty e^{at} x(t) e^{-st}\, dt = \int_0^\infty x(t) e^{-(s-a)t}\, dt$$

$$= \bar{x}(s - a)$$

since the integral is the Laplace transform of $x(t)$ with s replaced by $s - a$.

Example 7.6 Using (7.3), find $\mathcal{L}[t^2 e^{-t}]$. By (7.14), $\mathcal{L}[t^2 e^{-t}] = \bar{x}(s + 1)$, where $\bar{x}(s) = \mathcal{L}[t^2]$, and using (7.3) with $n = 2$, $\mathcal{L}[t^2] = 2/s^3$. Hence $\mathcal{L}[t^2 e^{-t}] = 2/(s + 1)^3$.

7.3.9 Second shifting theorem

$$\mathcal{L}[x(t-a)H(t-a)] = e^{-as}\mathcal{L}[x(t)] = e^{-as}\bar{x}(s) \qquad (7.15)$$

where $H(t-a)$ is the unit step function defined as

$$H(t-a) = \begin{cases} 0 & t<a \\ 1 & t>a \end{cases}$$

(Note that it is not defined for $t = a$).

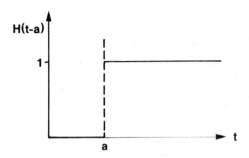

Fig. 7.1

If $x(t)$ is as plotted in Fig. 7.2 for $t \geqslant 0$, then the function $x(t-a)H(t-a)$ is as plotted in Fig. 7.3. It will be noticed that the curve is 'shifted' a distance a.

Proof

$$\mathcal{L}[x(t-a)H(t-a)] = \int_0^\infty x(t-a)H(t-a)e^{-st}\,dt$$

$$= \int_a^\infty x(t-a)e^{-st}\,dt$$

since the integrand is zero for $t < a$. On making the substitution $t - a = z$,

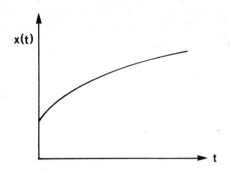

Fig. 7.2

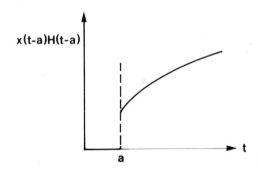

Fig. 7.3

this becomes $\mathcal{L}[x(t-a)H(t-a)] = \int\limits_{z=0}^{z=\infty} x(z)e^{-s(z+a)}\,dz$

$$= e^{-as}\int\limits_0^\infty x(z)e^{-sz}\,dz$$

$$= e^{-as}\bar{x}(s)$$

The unit step function will be discussed more fully in Section 7.4, when the use of (7.15) will be illustrated.

7.3.10 Periodic function

If $x(t)$ is a periodic function of period T, that is, if $x(t) = x(t + T)$, then

$$\mathcal{L}[x(t)] = \frac{1}{1 - e^{-sT}} \int_0^T e^{-st} x(t)\, dt \qquad (7.16)$$

Proof

$$\mathcal{L}[x(t)] = \int_0^\infty e^{-st} x(t)\, dt$$

$$= \int_0^T e^{-st} x(t)\, dt + \int_T^{2T} e^{-st} x(t)\, dt + \cdots \int_{nT}^{(n+1)T} e^{-st} x(t)\, dt + \cdots$$

Consider

$$\int_{nT}^{(n+1)T} e^{-st} x(t)\, dt$$

and let $t = nT + y$.

Then

$$\int_{nT}^{(n+1)T} e^{-st} x(t)\, dt = e^{-snT} \int_{y=0}^{y=T} e^{-sy} x(nT + y)\, dy$$

and since x is periodic, then $x(nT + y) = x(y)$, so

$$\int_{nT}^{(n+1)T} e^{-st} x(t)\, dt = e^{-snT} \int_0^T e^{-sy} x(y)\, dy = e^{-snT} \int_0^T e^{-st} x(t)\, dt$$

Hence

$$\mathcal{L}[x(t)] = [1 + e^{-sT} + e^{-2sT} + \cdots + e^{-nsT} + \cdots] \int_0^T e^{-st} x(t)\, dt$$

Noting that $s > 0$ for $\bar{x}(s)$ to exist, and hence that $0 < e^{-sT} < 1$, we can use the result

$$\frac{1}{1-x} = 1 + x + x^2 + \cdots + x^n + \cdots \qquad |x| < 1$$

to obtain (7.16).

Laplace transforms of some useful periodic functions are given in Barnett and Cronin (1975), Section 6.5.

Example 7.7 Determine the Laplace transform of the function $x(t)$ in Fig. 7.4 defined by

$$x(t) = \begin{cases} 1 & 0 \leqslant t < 1 \\ 0 & 1 \leqslant t < 2 \end{cases}$$

$$x(t) = x(t+2)$$

$$\mathcal{L}[x(t)] = \frac{1}{1-e^{-2s}} \int_0^2 x(t)e^{-st}\,dt = \frac{1}{1-e^{-2s}} \int_0^1 e^{-st}\,dt$$

since $x(t) = 0$ for $1 \leqslant t < 2$. Thus

$$\mathcal{L}[x(t)] = \frac{1}{1-e^{-2s}} \frac{1}{s}(1-e^{-s}) = \frac{1}{s(1+e^{-s})}$$

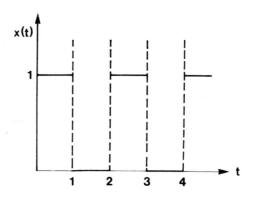

Fig. 7.4

7.3.11 Convolution theorem

$$\mathcal{L}\left[\int_0^t x(u)y(t-u)\,du\right] = \bar{x}(s)\bar{y}(s) \qquad (7.17)$$

This is used mostly when finding the inverse Laplace transform of the product of two functions. The proof involves the change of order of integration of a double integral, and is left as a problem for those familiar with this technique.

Problem 7.2 Find the Laplace transform of the following functions:

(a) $2e^{-3t}$
(b) $t^3 - te^{-t}$
(c) $5\cos 3t$
(d) $(t^3 + 1)^2$
(e) $e^t \sin 2t$

Problem 7.3 Sketch the following periodic functions and find their Laplace transform:

(a) $f(t) = (k/T)t$, $0 < t < T$, with $f(t+T) = f(t)$
(b) $f(t) = |\sin \omega t|$

7.4 The Heaviside unit step function

The Heaviside unit step function, as stated in Section 7.3.9, is defined by

$$H(t-a) = \begin{cases} 0 & t < a \\ 1 & a < t \end{cases}$$

A sketch is given in Fig. 7.1
It follows from (7.15) that

$$\mathcal{L}[H(t-a)] = e^{-as}\mathcal{L}[1] = e^{-as}/s$$

The Heaviside unit step function is useful in the description of functions which have discontinuous steps in their values at certain values of the independent variable. For example, if we consider a simple electrical circuit with an on–off switch which is switched on at time t_0, and such that the electromotive force provided by the battery is zero for $t < t_0$ and is the constant value E_0 for $t > t_0$, then the electromotive force $E(t)$ at any time t is given by $E(t) = E_0 H(t - t_0)$.

We now consider how a function which is given by different expressions in different ranges can be represented by a single expression using the Heaviside step function.

Let the function $f(t)$ be given by

$$f(t) = \begin{cases} f_0(t) & 0 < t < t_1 \\ f_1(t) & t_1 < t < t_2 \\ \vdots \\ f_r(t) & t_r < t < t_{r+1} \\ \vdots \\ f_n(t) & t_n < t \end{cases}$$

Rewrite the function as

$$f(t) = \begin{cases} f_0(t) & 0 < t < t_1 \\ f_0(t) + [f_1(t) - f_0(t)] & t_1 < t < t_2 \\ \vdots \\ f_{r-1}(t) + [f_r(t) - f_{r-1}(t)] & t_r < t < t_{r+1} \\ \vdots \\ f_{n-1}(t) + [f_n(t) - f_{n-1}(t)] & t_n < t \end{cases}$$

Then it is easily seen that $f(t)$ can be written as

$$\begin{aligned} f(t) = f_0(t) &+ [f_1(t) - f_0(t)]H(t - t_1) \\ &+ [f_2(t) - f_1(t)]H(t - t_2) + \cdots \\ &+ [f_r(t) - f_{r-1}(t)]H(t - t_r) + \cdots \\ &+ [f_n(t) - f_{n-1}(t)]H(t - t_n) \end{aligned} \quad (7.18)$$

since if we consider the general range

$$t_r < t < t_{r+1}$$

then

$$H(t - t_1) = H(t - t_2) = \cdots = H(t - t_r) = 1$$

and

$$H(t - t_{r+1}) = \cdots = H(t - t_n) = 0$$

so that

$$f(t) = f_0(t) + [f_1(t) - f_0(t)] + [f_2(t) - f_1(t)]$$
$$+ \cdots + [f_r(t) - f_{r-1}(t)] = f_r(t)$$

which is the required expression in the range.

Example 7.8 The function in Fig. 7.5 is given by

$$x(t) = \begin{cases} 0 & 0 < t < 1 \\ 2 & 1 < t < 2 \\ 0 & 2 < t \end{cases}$$

Rewrite as

$$x(t) = \begin{cases} 0 & 0 < t < 1 \\ 0 + (2 - 0) & 1 < t < 2 \\ 2 + (0 - 2) & 2 < t \end{cases}$$

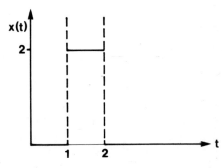

Fig. 7.5

So that

$$x(t) = 0 + (2 - 0)H(t - 0) + (0 - 2)H(t - 2)$$
$$= 2H(t - 1) - 2H(t - 2)$$

Hence

$$\mathcal{L}[x(t)] = e^{-s}\mathcal{L}[2] - e^{-2s}\mathcal{L}[2] \quad \text{(by 7.15)}$$
$$= \frac{2e^{-s}}{s} - \frac{2e^{-2s}}{s} = \frac{2e^{-s}}{s}(1 - e^{-s})$$

180   Ordinary Differential Equations

Example 7.9   The function in Fig. 7.6 is given by

$$x(t) = \begin{cases} 0 & 0 \leqslant t \leqslant 1 \\ t-1 & 1 \leqslant t \leqslant 3 \\ -2(t-4) & 3 \leqslant t \leqslant 4 \\ 0 & 4 \leqslant t \end{cases}$$

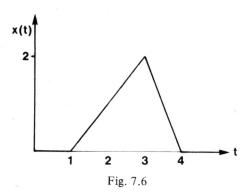

Fig. 7.6

Writing this as shown in (7.18), we have

$$x(t) = 0 + [(t-1) - 0]H(t-1) + [-2(t-4) - (t-1)]H(t-3)$$
$$+ [0 + 2(t-4)]H(t-4)$$
$$= (t-1)H(t-1) - 3(t-3)H(t-3) + 2(t-4)H(t-4)$$

On using (7.7) and the second shifting theorem, (7.15),

$$\mathcal{L}[x(t)] = e^{-s}\mathcal{L}[t] - 3e^{-3s}\mathcal{L}[t] + 2e^{-4s}\mathcal{L}[t]$$
$$= [e^{-s} - 3e^{-3s} + 2e^{-4s}]/s^2$$

Example 7.10   The function in Fig. 7.7 is given by

$$x(t) = \begin{cases} 2t & 0 \leqslant t < 1 \\ 1 & 1 < t < 2 \\ -2(t-3) & 2 < t \leqslant 3 \\ 0 & 3 \leqslant t \end{cases}$$

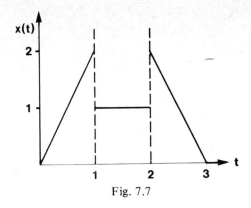

Fig. 7.7

Using (7.18),

$$x(t) = 2t + [1 - 2t]H(t - 1) + [-2(t - 3) - 1]H(t - 2)$$
$$+ [0 + 2(t - 3)]H(t - 3)$$
$$= 2t + (1 - 2t)H(t - 1) + (-2t + 5)H(t - 2) + 2(t - 3)H(t - 3)$$

The Laplace transform of a function $x(t - a)H(t - a)$ can be found using the second shifting theorem, (7.15), but in order to find the Laplace transform of $f(t)H(t - a)$, it is necessary to rewrite the function in the form $g(t - a)H(t - a)$. In this example it is therefore necessary to write $(1 - 2t)$ in terms of $(t - 1)$ and to write $(-2t + 5)$ in terms of $(t - 2)$.

That is,
$$(1 - 2t) = -2(t - \tfrac{1}{2}) = -2(t - 1 + \tfrac{1}{2}) = -2(t - 1) - 1$$
and
$$(5 - 2t) = -2(t - \tfrac{5}{2}) = -2(t - 2 - \tfrac{1}{2}) = -2(t - 2) + 1$$

Hence
$$x(t) = 2t - 2(t - 1)H(t - 1) - H(t - 1) - 2(t - 2)H(t - 2)$$
$$+ H(t - 2) + 2(t - 3)H(t - 3)$$

and
$$\bar{x}(s) = \frac{2}{s^2} - \frac{2e^{-s}}{s^2} - \frac{e^{-s}}{s} - \frac{2e^{-2s}}{s^2} + \frac{e^{-2s}}{s} + \frac{2e^{-3s}}{s^2}$$
$$= [2 - e^{-s}(2 + s) + e^{-2s}(s - 2) + 2e^{-3s}]/s^2$$

Example 7.11 Consider the function $x(t)$ defined by

$$x(t) = \begin{cases} 0 & 0 \leq t \leq \pi/2 \\ -\cos t & \pi/2 \leq t \leq \pi \\ 0 & \pi < t \end{cases}$$

Then $x(t)$ can be expressed as

$$x(t) = -\cos t\, H(t - \pi/2) + \cos t\, H(t - \pi),$$
$$= -\cos[(t - \pi/2) + \pi/2]H(t - \pi/2) + \cos[(t - \pi) + \pi]H(t - \pi)$$
$$= \sin(t - \pi/2)H(t - \pi/2) - \cos(t - \pi)H(t - \pi)$$

Hence

$$\mathcal{L}[x(t)] = e^{-(\pi/2)s}\frac{1}{s^2+1} - e^{-\pi s}\frac{s}{s^2+1} = \frac{e^{-(\pi/2)s}}{s^2+1}[1 - se^{-(\pi/2)s}]$$

Problem 7.4 Express the following functions using the Heaviside step function, and hence find their Laplace transforms:

(a) $f(t) = \begin{cases} t & 0 < t < 1 \\ 1 & 1 < t \end{cases}$

(b) $f(t) = \begin{cases} |\sin t| & 0 < t < 2\pi \\ 0 & 2\pi < t \end{cases}$

(c) $f(t) = \begin{cases} t & 0 < t < a \\ 2a - t & a < t < 2a \end{cases}$
$f(t + 2a) = f(t)$

(d) $f(t) = \begin{cases} te^{-t} & 0 < t < 2 \\ 1 & 2 < t \end{cases}$

7.5 The delta function

The delta function $\delta(t - a)$ is defined as

$$\delta(t - a) = \begin{cases} 0 & t < a \\ 0 & t > a \end{cases}$$

and

$$\int_{-\infty}^{+\infty} \delta(t - a)\, dt = 1$$

It follows that the only non-zero value for $\delta(t - a)$ is at $t = a$. The delta function can be described by considering Fig. 7.8.
  The function $x(t)$ is taken to be

$$x(t) = \begin{cases} 0 & t < a \\ 1/\epsilon & a < t < a + \epsilon \\ 0 & a + \epsilon < t \end{cases} \qquad (7.19)$$

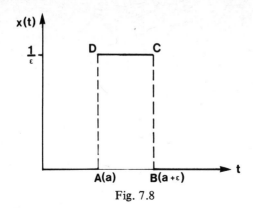

Fig. 7.8

The area of the rectangle $ABCD$ is unity, that is, $\int_{-\infty}^{\infty} x(t)\,dt = 1$, whatever the value of ϵ. If now $\epsilon \to 0$, then $x(t) \to \delta(t-a)$.

From the definition of the delta function, it follows that

$$\int_{-\infty}^{\infty} f(t)\delta(t-a)\,dt = f(a)$$

provided that the integral exists and that $f(t)$ is continuous in the neighbourhood of $t = a$. Hence

$$\mathcal{L}[\delta(t-a)] = \int_0^{\infty} e^{-st}\delta(t-a)\,dt = e^{-sa} \qquad a \geqslant 0 \qquad (7.20)$$

We will now also obtain $\mathcal{L}[\delta(t-a)]$ by taking the Laplace transform of (7.19) and letting $\epsilon \to 0$.

Now $x(t)$ as defined by (7.19) can also be written as

$$x(t) = \frac{1}{\epsilon}[H(t-a) - H(t-a-\epsilon)]$$

Hence

$$\mathcal{L}[x(t)] = \frac{1}{\epsilon}[e^{-as} - e^{-(a+\epsilon)s}] = e^{-as}\left[\frac{1-e^{-\epsilon s}}{\epsilon}\right]$$

Hence

$$\mathcal{L}[\delta(t-a)] = e^{-as} \lim_{\epsilon \to 0} (1-e^{-\epsilon s})/\epsilon = e^{-as}$$

since it can be shown, for instance using L'Hospital's rule, that

$$\lim_{\epsilon \to 0} (1-e^{-\epsilon s})/\epsilon = 1$$

We have as a special case

$$\mathcal{L}[\delta(t)] = 1 \qquad (7.21)$$

The delta function is useful in dealing with physical situations involving impulses, such as a lightning stroke on a transmission line or a hammer blow on a mechanical system. An example is given in Example 7.18.

7.6 The inverse Laplace transform

7.6.1 Definition

If the Laplace transform of the function $x(t)$ is $\bar{x}(s)$, then $x(t)$ is called the *inverse Laplace transform* of $\bar{x}(s)$. It follows from the definition of the Laplace transform that the inverse is valid only for $t \geqslant 0$.

The inverse is written as

$$\mathcal{L}^{-1}[\bar{x}(s)] = x(t) \qquad (7.22)$$

We have seen from eq. (7.5) that

$$\mathcal{L}[\sin t] = \frac{1}{(s^2 + 1)}$$

and it follows therefore that

$$\mathcal{L}^{-1}\left[\frac{1}{(s^2 + 1)}\right] = \sin t$$

A theorem states that under fairly general conditions the inverse is unique and we shall always assume uniqueness.

In this book the inverse Laplace transform will be obtained by the use of tables of transforms. However, the methods of complex-variable theory can be used to find the inverse transform, using the so-called complex inversion integral.

7.6.2 Rules for inversion

All the rules listed in Section 7.3 are valid for inverse Laplace transforms. For example, (7.14) can be read as

$$\mathcal{L}^{-1}[\bar{x}(s-a)] = e^{at}\mathcal{L}^{-1}[\bar{x}(s)] = e^{at}x(t)$$

and (7.15) as
$$\mathcal{L}^{-1}[e^{-as}\bar{x}(s)] = x(t-a)H(t-a)$$
where
$$x(t) = \mathcal{L}^{-1}[\bar{x}(s)]$$

The evaluation of the inverse Laplace transform and the use of the rules will be illustrated by a number of examples. The rules henceforth will be referred to the list given in Table A of the appendix and such that A8 refers to rule 8 of Table A, etc. Similarly B6 refers to the transform pair number 6 from Table B.

Example 7.12 Find
$$\mathcal{L}^{-1}\left[\frac{s}{s^2 + 2s + 2}\right]$$

It is necessary to complete the square of the denominator, that is
$$s^2 + 2s + 2 = (s+1)^2 + 1$$

Hence
$$\frac{s}{s^2 + 2s + 2} = \frac{s}{(s+1)^2 + 1} = \frac{s+1}{(s+1)^2 + 1} - \frac{1}{(s+1)^2 + 1}$$

and so by A1
$$\mathcal{L}^{-1}\left[\frac{s}{s^2 + 2s + 2}\right] = \mathcal{L}^{-1}\left[\frac{s+1}{(s+1)^2 + 1}\right] - \mathcal{L}^{-1}\left[\frac{1}{(s+1)^2 + 1}\right]$$

The two terms on the right-hand side are functions of $(s+1)$, and so, by A8, with a replaced by -1,
$$\mathcal{L}^{-1}\left[\frac{s}{s^2 + 2s + 2}\right] = e^{-t}\mathcal{L}^{-1}\left[\frac{s}{s^2 + 1}\right] - e^{-t}\mathcal{L}^{-1}\left[\frac{1}{s^2 + 1}\right]$$

Hence from tables of inverse Laplace transforms, see B6 and B7, we have
$$\mathcal{L}^{-1}\left[\frac{s}{s^2 + 2s + 2}\right] = e^{-t}\cos t - e^{-t}\sin t$$

Example 7.13 Find
$$\mathcal{L}^{-1}\left[\frac{1}{s(s^2+4)}\right]$$

This inverse can be obtained in several different ways.

(a) Express the function in partial fractions:
$$\frac{1}{s(s^2+4)} = \frac{1}{4s} - \frac{s}{4(s^2+4)}$$

Hence
$$\mathcal{L}^{-1}\left[\frac{1}{s(s^2+4)}\right] = \tfrac{1}{4}\mathcal{L}^{-1}\left[\frac{1}{s}\right] - \tfrac{1}{4}\mathcal{L}^{-1}\left[\frac{s}{s^2+4}\right]$$
$$= \tfrac{1}{4}(1 - \cos 2t), \text{ using B1 and B7.}$$

(b) By use of A4,
$$\mathcal{L}^{-1}\left[\frac{1}{s(s^2+4)}\right] = \int_0^t x(u)\, du$$

where
$$x(t) = \mathcal{L}^{-1}\left[\frac{1}{s^2+4}\right] = \tfrac{1}{2}\mathcal{L}^{-1}\left[\frac{2}{s^2+2^2}\right] = \tfrac{1}{2}\sin 2t \qquad \text{(using B6)}$$

Hence
$$\mathcal{L}^{-1}\left[\frac{1}{s(s^2+4)}\right] = \tfrac{1}{2}\int_0^t \sin 2u\, du = \tfrac{1}{4}(1 - \cos 2t)$$

(c) By use of the convolution theorem A11, with
$$x(t) = \mathcal{L}^{-1}\left[\frac{1}{s}\right] = 1 \quad \text{and} \quad y(t) = \mathcal{L}^{-1}\left[\frac{1}{(s^2+4)}\right] = \tfrac{1}{2}\sin 2t$$

$$\mathcal{L}^{-1}\left[\frac{1}{s(s^2+4)}\right] = \int_0^t 1 \cdot \tfrac{1}{2}\sin 2(t-u)\, du = \tfrac{1}{4}(1 - \cos 2t)$$

Example 7.14 Find
$$\mathcal{L}^{-1}\left[\frac{s}{(s^2+a^2)^2}\right]$$

(a) Since

$$\frac{d}{ds}\left(\frac{1}{s^2+a^2}\right) = -\frac{2s}{(s^2+a^2)^2}$$

then

$$\mathcal{L}^{-1}\left[\frac{s}{(s^2+a^2)^2}\right] = -\tfrac{1}{2}\mathcal{L}^{-1}\left[\frac{d}{ds}\left(\frac{1}{s^2+a^2}\right)\right]$$

$$= -\tfrac{1}{2}(-t)\mathcal{L}^{-1}\left[\frac{1}{s^2+a^2}\right] \quad \text{(by A5)}$$

$$= \frac{t}{2a}\sin at$$

(b) By use of the convolution theorem A11, with

$$x(t) = \mathcal{L}^{-1}\left[\frac{1}{(s^2+a^2)}\right] = \frac{1}{a}\sin at$$

and

$$y(t) = \mathcal{L}^{-1}\left[\frac{s}{(s^2+a^2)}\right] = \cos at$$

we have

$$\mathcal{L}^{-1}\left[\frac{s}{(s^2+a^2)^2}\right] = \int_0^t \frac{1}{a}\sin au \cos a(t-u)\,du$$

$$= \frac{1}{2a}\int_0^t [\sin at + \sin a(2u-t)]\,du$$

$$= \frac{1}{2a}\left[u\sin at - \frac{1}{2a}\cos a(2u-t)\right]_0^t$$

$$= \frac{1}{2a}t\sin at$$

188 Ordinary Differential Equations

Notice that it is immaterial whether $x(t)$ is chosen as above or as

$$\mathcal{L}^{-1}\left[\frac{s}{(s^2+a^2)}\right]$$

with $y(t)$ as

$$\mathcal{L}^{-1}\left[\frac{1}{(s^2+a^2)}\right]$$

as in this case we obtain

$$\mathcal{L}^{-1}\left[\frac{s}{(s^2+a^2)^2}\right] = \int_0^t (\cos au)\frac{1}{a}\sin a(t-u)\,du$$

which can be integrated in a similar manner to that above, and which gives the same result. For some functions, however, one of the integrals may be more readily evaluated than the other, so that some thought given to the choice of $x(t)$ and $y(t)$ is advisable.

Example 7.15 Find

$$\mathcal{L}^{-1}\left[\frac{e^{-s}}{s^2(s+1)}\right]$$

(a) By the use of partial fractions,

$$\frac{e^{-s}}{s^2(s+1)} = \left[\frac{1}{s^2} - \frac{1}{s} + \frac{1}{s+1}\right]e^{-s}$$

and

$$\mathcal{L}^{-1}\left[\frac{1}{s^2} - \frac{1}{s} + \frac{1}{s+1}\right] = t - 1 + e^{-t}$$

Hence by A9, with $a = 1$

$$\mathcal{L}^{-1}\left[\frac{e^{-s}}{s^2(s+1)}\right] = [(t-1) - 1 + e^{-(t-1)}]H(t-1)$$

(b) By use of the convolution theorem A11 with

$$x(t) = \mathcal{L}^{-1}\left[\frac{e^{-s}}{s^2}\right] = (t-1)H(t-1)$$

and

$$y(t) = \mathcal{L}^{-1}\left[\frac{1}{s+1}\right] = e^{-t}$$

we have

$$\mathcal{L}^{-1}\left[\frac{e^{-s}}{s^2(s+1)}\right] = \int_0^t (u-1)H(u-1)e^{-(t-u)}\,du$$

If $t < 1$, it follows that $H(u-1) = 0$ for all values of the variable of integration u, since $0 < u < t$. Hence the integral is zero, that is

$$\mathcal{L}^{-1}\left[\frac{e^{-s}}{s^2(s+1)}\right] = 0 \qquad t < 1$$

If $t > 1$, the integral can be written as

$$\int_0^1 (u-1)H(u-1)e^{-(t-u)}\,du + \int_1^t (u-1)H(u-1)e^{-(t-u)}\,du$$

In the first integral, $H(u-1) = 0$ since $0 < u < 1$, and in the second integral $H(u-1) = 1$ since $u > 1$. Hence, for $t > 1$,

$$\mathcal{L}^{-1}\left[\frac{e^{-s}}{s^2(s+1)}\right] = \int_1^t (u-1)e^{-(t-u)}\,du = e^{-t}\int_1^t (u-1)e^u\,du$$

$$= e^{-t}[e^u(u-2)]_1^t = (t-2) + e^{-(t-1)}$$

The inverse for $t < 1$ can be combined with that for $t > 1$ to give the single expression, valid for all t,

$$\mathcal{L}^{-1}\left[\frac{e^{-s}}{s^2(s+1)}\right] = [(t-2) + e^{-(t-1)}]H(t-1)$$

Problem 7.5 Find the inverse Laplace transform of each of the following functions:

(a) $\dfrac{1}{s^2+4}$ \qquad (b) $\dfrac{3}{s+3}$

(c) $\dfrac{s}{s^2+3}$ \qquad (d) $\dfrac{s+1}{s^3} - \dfrac{s-1}{s^2+9}$

(e) $\dfrac{s-1}{s^2+4s+9}$ \qquad (f) $\dfrac{s}{s^2-s+3}$

(g) $\dfrac{1}{s^2(s+1)}$ \qquad (h) $\dfrac{s}{(s+1)^2(s+2)}$

190   Ordinary Differential Equations

(i) $\dfrac{s+1}{(s^2+2s+3)(s-1)}$

(j) $\dfrac{1}{(s^2+4s+5)(s^2+4s+4)}$

(k) $\dfrac{s}{(s^2+1)^3}$

(l) $\dfrac{1}{(s-1)^{3/2}}$

(m) $\dfrac{e^{-s}}{s}$

(n) $\dfrac{e^{-2s}}{(s+1)^2}$

(o) $\dfrac{e^{-\pi s}}{s^2+2s+2}$

(p) $\dfrac{e^{-s}}{s^2(s+1)}$

7.7 Solution of linear differential equations

The method of solution has been described at the beginning of this chapter, so that the use of the Laplace transform is shown by a number of examples.

Example 7.16   Find the general solution of

$$x'' + 4x' + 4x = e^{-2t}$$

The initial conditions are not known, so let $x = A$ at $t = 0$ and let $dx/dt = B$ at $t = 0$. Then on taking the Laplace transform, we obtain

$$[s^2 \bar{x}(s) - As - B] + 4[s\bar{x}(s) - A] + 4\bar{x}(s) = \mathcal{L}[e^{-2t}] \qquad \text{(by A7)}$$

so that

$$\bar{x}(s)[s^2 + 4s + 4] - As - (B + 4A) = 1/(s+2) \qquad \text{(by B5)}$$

Solving for $\bar{x}(s)$,

$$\bar{x}(s) = \dfrac{As + (B + 4A)}{(s+2)^2} + \dfrac{1}{(s+2)^3}$$

$$= \dfrac{A(s+2) + (B + 2A)}{(s+2)^2} + \dfrac{1}{(s+2)^3}$$

$$= \dfrac{A}{(s+2)} + \dfrac{B + 2A}{(s+2)^2} + \dfrac{1}{(s+2)^3}$$

On taking the inverse, we obtain

$$x(t) = e^{-2t}[A + (B + 2A)t + \tfrac{1}{2}t^2] \qquad \text{(by A8 and B3)}$$

and this is the general solution since it involves two unknown constants.

Example 7.17 Solve the pair of simultaneous differential equations
$$dx/dt = x - y \qquad dy/dt = -2x$$
where $x = 2$ and $y = 1$ when $t = 0$.

Taking the Laplace transform of the pair of equations and using the initial conditions with A3, we get
$$s\bar{x}(s) - 2 = \bar{x}(s) - \bar{y}(s)$$
$$s\bar{y}(s) - 1 = -2\bar{x}(s)$$

Rearranging the terms to bring the unknown functions $\bar{x}(s)$ and $\bar{y}(s)$ to the left-hand side of the equations, we have
$$(s - 1)\bar{x}(s) + \bar{y}(s) = 2$$
$$2\bar{x}(s) + s\bar{y}(s) = 1$$

This pair of simultaneous algebraic equations can be easily solved by elimination, Cramer's rule or some other method to give
$$\bar{x}(s) = \frac{2s - 1}{s^2 - s - 2} \qquad \text{and} \qquad \bar{y}(s) = \frac{s - 5}{s^2 - s - 2}$$

Hence
$$x(t) = \mathcal{L}^{-1}\left[\frac{2s - 1}{(s + 1)(s - 2)}\right] = \mathcal{L}^{-1}\left[\frac{1}{s + 1} + \frac{1}{s - 2}\right]$$
$$= e^{-t} + e^{2t} \qquad \text{(using B5)}$$

and
$$y(t) = \mathcal{L}^{-1}\left[\frac{s - 5}{(s + 1)(s - 2)}\right] = \mathcal{L}^{-1}\left[\frac{2}{s + 1} - \frac{1}{s - 2}\right]$$
$$= 2e^{-t} - e^{2t}$$

Example 7.18 The voltage V of a certain circuit satisfies
$$V'' + 5V' + 4V = E(t)$$
where E is the applied voltage. Find the voltage in the circuit when E

192 Ordinary Differential Equations

is a unit impulse at $t = 0$, and if initially the voltage and the rate of change of voltage are zero.

We have that $E(t) = \delta(t)$ and that $V = dV/dt = 0$ when $t = 0$. Taking the Laplace transform of the differential equation, using B7 and A14,

$$s^2 \bar{V}(s) + 5s\bar{V}(s) + 4\bar{V}(s) = \mathcal{L}[\delta(t)] = 1$$

Hence

$$\bar{V}(s) = \frac{1}{(s+1)(s+4)} = \tfrac{1}{3}\left[\frac{1}{s+1} - \frac{1}{s+4}\right]$$

Hence

$$V(t) = \tfrac{1}{3}(e^{-t} - e^{-4t}) \tag{7.23}$$

We note that

$$dV/dt = \tfrac{1}{3}(4e^{-4t} - e^{-t})$$

and hence that $dV/dt = 1$ when $t = 0$. This does not agree with the given initial condition.

To illustrate what has happened, we take

$$E(t) = \frac{1}{\epsilon}[1 - H(t - \epsilon)] \qquad \epsilon \to 0$$

With $E(t)$ defined in this way,

$$(s+1)(s+4)\bar{V}(s) = [1 - e^{-\epsilon s}]/\epsilon s$$

Hence, on inversion,

$$12\epsilon V(t) = e^{-4t} - 4e^{-t} + 3 - [e^{-4(t-\epsilon)} - 4e^{-(t-\epsilon)} + 3]H(t - \epsilon)$$

That is,

$$12\epsilon V(t) = \begin{cases} e^{-4t} - 4e^{-t} + 3 & 0 \leqslant t \leqslant \epsilon \\ (e^{-4t} - 4e^{-t} + 3) - [e^{-4(t-\epsilon)} - 4e^{-(t-\epsilon)} + 3] & \epsilon \leqslant t \end{cases}$$

and

$$3\epsilon \, dV/dt = \begin{cases} e^{-t} - e^{-4t} & 0 \leqslant t \leqslant \epsilon \\ (e^{-t} - e^{-4t}) - [e^{-(t-\epsilon)} - e^{-4(t-\epsilon)}] & \epsilon \leqslant t \end{cases}$$

Both the functions are continuous (Fig. 7.9) and with

$$V(0) = V'(0) = 0$$

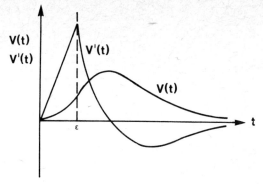

Fig. 7.9

and
$$V(\epsilon) = (e^{-4\epsilon} - 4e^{-\epsilon} + 3)/12\epsilon \simeq \tfrac{1}{2}\epsilon$$
$$V'(\epsilon) = (e^{-\epsilon} - e^{-4\epsilon})/3\epsilon \simeq 1$$

If we now let $\epsilon \to 0$, we obtain
$$V(t) = \tfrac{1}{3}(e^{-t} - e^{-4t}) \qquad 0 \leqslant t$$

and
$$V'(t) = \begin{cases} 0 & t = 0 \\ \tfrac{1}{3}(4e^{-4t} - e^{-t}) & 0 < t \end{cases}$$

The expression for the voltage agrees with that given by (7.23), and the expression for the rate of change of voltage illustrates that the effect of the impulse at $t = 0$ is to cause a discontinuous jump of magnitude unity in $V'(t)$ at $t = 0$.

Example 7.19 The equation of motion of a particle moving along the z axis is $z'' + z = 6 \cos 2t$, where z is the distance from the origin in time t. Determine z as a function of t if the particle is at the origin in time $t = 0$ when the velocity is unity in the positive z direction.

The Laplace transform of the differential equation is, using A7,
$$[s^2 \bar{z}(s) - sz(0) - z'(0)] + \bar{z}(s) = \mathcal{L}[6 \cos 2t]$$

But for this motion, $z(0) = 0$ and $z'(0) = 1$; hence
$$(s^2 + 1)\bar{z}(s) - 1 = \mathcal{L}[6 \cos 2t] = \frac{6s}{s^2 + 4} \qquad \text{(using B7)}$$

On solving this algebraic equation, we get

$$\bar{z}(s) = \frac{1}{s^2 + 1}\left[1 + \frac{6s}{s^2 + 4}\right]$$

Taking the inverse transform:

$$z(t) = \mathcal{L}^{-1}\left[\frac{1}{s^2 + 1}\right] + 6\mathcal{L}^{-1}\left[\frac{s}{(s^2 + 1)(s^2 + 4)}\right]$$

The first term can be obtained from B6.
 To find the inverse of

$$\frac{s}{(s^2 + 1)(s^2 + 4)}$$

the function can be put in terms of its partial fractions

$$\left[\frac{As + B}{s^2 + 1} + \frac{Cs + D}{s^2 + 4}\right]$$

or the convolution theorem can be used. We will use the convolution theorem, A11, with

$$x(t) = \mathcal{L}^{-1}[s/(s^2 + 4)] \quad \text{and} \quad y(t) = \mathcal{L}^{-1}[1/(s^2 + 1)]$$

Hence

$$6\mathcal{L}^{-1}\left[\frac{s}{(s^2 + 1)(s^2 + 4)}\right] = 6\int_0^t \cos 2u \sin(t - u)\, du$$

$$= 3\int_0^t [\sin(t + u) + \sin(t - 3u)]\, du$$

$$= [-3\cos(t + u) + \cos(t - 3u)]_0^t$$

$$= 2(\cos t - \cos 2t)$$

so that $z(t) = \sin t + 2(\cos t - \cos 2t)$.
 Note that if the convolution theorem is used, it is not necessary to evaluate the Laplace transform of the right-hand side of the differential equation (the 'forcing function'). For this example,

$$\bar{x}(s) = \frac{1}{s^2 + 1}\{1 + \mathcal{L}[6\cos 2t]\}$$

so

$$x(t) = \sin t + \mathcal{L}^{-1}\left\{\frac{1}{s^2 + 1}\mathcal{L}[6\cos 2t]\right\}$$

If, in the above equation, $x(t)$ is taken as
$$\mathcal{L}^{-1}[\mathcal{L}(t \cos 2t)] = 6 \cos 2t$$
and $y(t)$ as $\mathcal{L}^{-1}(1/(s^2 + 1)) = \sin t$, then
$$\mathcal{L}^{-1}\left\{\frac{1}{s^2 + 1} \mathcal{L}[6 \cos 2t]\right\} = \int_0^t 6 \cos 2u \sin(t - u) \, du$$
as before.

A similar procedure can be applied for all linear differential equations.

Example 7.20 Find the charge $Q(t)$ in the electrical circuit shown in Fig. 7.10 if the electromotive force $e(t)$ provided by the battery is given by
$$e(t) = \begin{cases} 8t \text{ V} & 0 < t < 2 \text{ s} \\ 0 & t > 2 \text{ s} \end{cases}$$
and if initially the charge and current are zero.

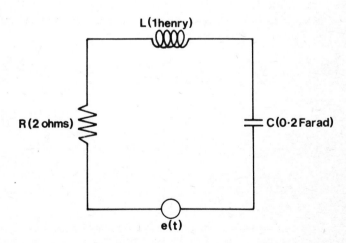

Fig. 7.10

196 Ordinary Differential Equations

The voltage drop across the resistance R is
$$Ri = 2\frac{dQ}{dt}$$
that across the induction L is
$$L\frac{di}{dt} = \frac{d^2Q}{dt^2}$$
that across the capacitor C is
$$\frac{Q}{C} = 5Q$$
that across the battery is $-e$.

The total voltage drop around the closed circuit is zero. Hence the differential equation which gives the charge is
$$Q'' + 2Q' + 5Q = e(t)$$
with initial conditions $Q(0) = 0$, $i(0) = Q'(0) = 0$.
Taking the Laplace transform, we obtain
$$s^2\bar{Q}(s) + 2s\bar{Q}(s) + 5\bar{Q}(s) = \mathcal{L}[e(t)]$$
hence
$$\bar{Q}(s) = \frac{1}{s^2 + 2s + 5}\mathcal{L}[e(t)]$$

The inverse transform will be obtained in two ways, firstly using partial fractions and secondly using the convolution theorem and bypassing the evaluation of $\mathcal{L}[e(t)]$. Which method the reader chooses in any particular problem depends mainly on personal preference.

(a) Partial fractions
$$\mathcal{L}[e(t)] = \mathcal{L}[8t - 8tH(t-2)]$$
$$= \mathcal{L}[8t - 16H(t-2) - 8(t-2)H(t-2)]$$
$$= 8\left[\frac{1}{s^2} - \frac{2e^{-2s}}{s} - \frac{e^{-2s}}{s^2}\right]$$
Hence
$$Q(t) = 8\mathcal{L}^{-1}\left\{\frac{1}{s^2 + 2s + 5}\left[\frac{1}{s^2} - \frac{2e^{-2s}}{s} - \frac{e^{-2s}}{s^2}\right]\right\}$$
$$= \frac{8}{25}\mathcal{L}^{-1}\left[\left(-\frac{2}{s} + \frac{5}{s^2} + \frac{2s-1}{s^2 + 2s + 5}\right)(1 - e^{-2s})\right.$$
$$\left. - 10\left(\frac{1}{s} - \frac{s+2}{s^2 + 2s + 5}\right)e^{-2s}\right]$$

on converting

$$\frac{1}{s^2(s^2+2s+5)} \quad \text{and} \quad \frac{1}{s(s^2+2s+5)}$$

into partial fractions.
Hence

$$Q(t) = \tfrac{8}{25}\mathcal{L}^{-1}\left\{\left[-\frac{2}{s}+\frac{5}{s^2}+\frac{2(s+1)}{(s+1)^2+4}-\frac{3}{(s+1)^2+4}\right]\right.$$

$$\left.+ e^{-2s}\left[-\frac{8}{s}-\frac{5}{s^2}+\frac{8(s+1)}{(s+1)^2+4}+\frac{13}{(s+1)^2+4}\right]\right\}$$

$$= \tfrac{8}{25}[-2+5t+(2\cos 2t-\tfrac{3}{2}\sin 2t)e^{-t}]$$
$$+ \tfrac{8}{25}\{-8-5(t-2)+[8\cos 2(t-2)$$
$$+ \tfrac{13}{2}\sin 2(t-2)]e^{-(t-2)}\}H(t-2)$$

(b) With

$$x(t) = e(t) \quad \text{and} \quad y(t) = \mathcal{L}^{-1}\left[\frac{1}{s^2+2s+5}\right] = \tfrac{1}{2}e^{-t}\sin 2t$$

in A11 we have

$$\mathcal{L}^{-1}\left\{\frac{1}{s^2+2s+5}\mathcal{L}[e(t)]\right\} = \int_0^t e(u)\tfrac{1}{2}e^{-(t-u)}\sin 2(t-u)\,du$$

$$= \tfrac{1}{2}e^{-t}\int_0^t e(u)\,e^u\,\sin 2(t-u)\,du$$

For $t<2$,

$$Q(t) = \tfrac{1}{2}e^{-t}\int_0^t 8ue^u\sin 2(t-u)\,du$$

which on evaluating the integral gives

$$Q(t) = \tfrac{8}{25}(5t-2)+\tfrac{4}{25}(4\cos 2t-3\sin 2t)e^{-t}$$

For $t>2$,

$$Q(t) = \tfrac{1}{2}e^{-t}\int_0^2 8ue^u\sin 2(t-u)\,du$$

since $e(u) = 0, u > 2$.
On evaluating the integral,

$$Q(t) = \tfrac{4}{25}[13\sin 2(t-2)+16\cos 2(t-2)]e^{-(t-2)}$$
$$+ \tfrac{4}{25}(4\cos 2t-3\sin 2t)e^{-t}$$

The reader should verify that the expression for $Q(t)$ given by (b) agrees with that given by (a).

Example 7.21 A horizontal beam of length l m is fixed at one end and freely supported at the other. The beam carries a uniform load of w N/m of length and a weight W N at the middle. Find the equation giving the small deformation of the beam.

Take axes as shown in Fig. 7.11. Let F be the unknown reaction of the support at $x = l$. Then the differential equation which governs small deformation of the beam is

$$EI\, d^2 y/dx^2 = M$$

where E is the constant modulus of elasticity,
I is the constant moment of inertia of the section of the beam,
M is the moment of the forces about P acting on either of the two segments into which P divides the beam.

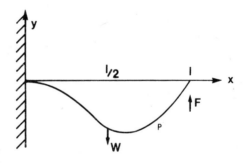

Fig. 7.11

M is taken as positive if in the anticlockwise sense.

If P is a distance x along the beam, then the moment of the forces acting on that section of the beam to the right of P is
for $\dfrac{l}{2} < x < l$

$$M = F(l - x) - \tfrac{1}{2} w(l - x)^2$$

(since the weight of the section is $w(l - x)$ and can be considered to act at the mid point of the section),

for $0 < x < \dfrac{l}{2}$

$$M = F(l - x) - \tfrac{1}{2}w(l - x)^2 - W(\tfrac{1}{2}l - x)$$

The moment can be written as one expression over the range $0 < x < l$, using the Heaviside step function, and is

$$M = F(l - x) - \tfrac{1}{2}w(l - x)^2 + W(x - \tfrac{1}{2}l)[1 - H(x - \tfrac{1}{2}l)]$$

The solution of the differential equation with M given above is needed subject to the boundary conditions

(a) $y = 0$ at $x = 0$

(b) $dy/dx = 0$ at $x = 0$

(c) $y = 0$ at $x = l$

These three conditions are sufficient to determine the two constants of integration and the reaction F. Taking the Laplace transform of the differential equation (noting that the independent variable is x instead of t), and using the two boundary conditions at $x = 0$, the transformed equation is

$$EIs^2\bar{y} = F\left(\frac{l}{s} - \frac{1}{s^2}\right) - \tfrac{1}{2}w\left(\frac{l^2}{s} - \frac{2l}{s^2} + \frac{2}{s^3}\right) + W\left(\frac{1}{s^2} - \frac{l}{2s}\right) - \frac{W}{s^2}e^{-ls/2}$$

Hence dividing by s^2 and taking the inverse Laplace transform:

$$EIy = F\left(\frac{lx^2}{2} - \frac{x^3}{6}\right) - \tfrac{1}{2}w\left(\frac{l^2 x^2}{2} - \frac{lx^3}{3} + \frac{x^4}{12}\right)$$

$$+ W\left(\frac{x^3}{6} - \frac{lx^2}{4}\right) - \frac{W}{6}\left(x - \frac{l}{2}\right)^3 H\left(x - \frac{l}{2}\right)$$

The unknown constant F can now be obtained from the remaining condition $y = 0$ at $x = l$, so finally

$$EIy = \frac{lx^2}{96}(6w + 5W)(3l - x) - \frac{W}{24}x^2(6l^2 - 4lx + x^2)$$

$$+ \frac{Wx^2}{12}(2x - 3l) - \frac{W}{48}(2x - l)^3 H\left(x - \frac{l}{2}\right)$$

200   Ordinary Differential Equations

Example 7.22   5000 kg/h of sulphuric acid, of specific heat 0·36, is cooled in a two stage cooler consisting of two tanks both with a capacity of 5000 kg. In the steady-state condition, hot acid at 160°C is fed to the first tank, where it is well stirred in contact with cooling coils. The acid leaves this tank at a temperature of 80°C and enters the second tank, where a similar cooling process takes place, the acid leaving at a temperature of 40°C. Find the temperature of the acid leaving both tanks one hour after the cooling is stopped.

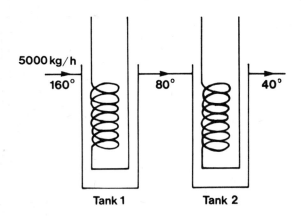

Fig. 7.12

Steady state. When cooling is stopped, the heat input minus the heat output must equal the heat accumulation in each tank.

At time t (hours) after cooling has stopped

let T_1 be the temperature of the acid leaving tank 1, and
   T_2 be the temperature of the acid leaving tank 2.

Tank 1
   Heat input   = (flow rate) × (specific heat) × (temperature)
                   = 5000 × 0·36 × 160
   Heat output   = 5000 × 0·36 × T_1

Heat accumulation = (Capacity of tank) × (specific heat) ×
(rate of change of temperature)
= 5000 × 0·36 × dT_1/dt

(Since the tank is well stirred, the temperature of acid in the tank can be taken to be that leaving the tank.)
Hence the heat balance is

$$5000 \times 0\cdot36 \times 160 - 5000 \times 0\cdot36 \times T_1 = 5000 \times 0\cdot36 \times \frac{dT_1}{dt}$$

that is,

$$160 - T_1 = \frac{dT_1}{dt}$$

Tank 2
Similarly

$$5000 \times 0\cdot36 \times T_1 - 5000 \times 0\cdot36 \times T_2 = 5000 \times 0\cdot36 \times \frac{dT_2}{dt}$$

that is,

$$T_1 - T_2 = \frac{dT_2}{dt}$$

We now have a pair of simultaneous differential equations with the initial conditions

$$T_1 = 80° \quad T_2 = 40° \quad \text{when } t = 0$$

Take the Laplace transforms of the equations

$$s\bar{T}_1 - 80 + \bar{T}_1 = 160/s$$
$$s\bar{T}_2 - 40 + \bar{T}_2 - \bar{T}_1 = 0$$

Solving, we get

$$\bar{T}_1 = \frac{160}{s(s+1)} + \frac{80}{s+1}$$

and

$$\bar{T}_2 = \frac{160}{s(s+1)^2} + \frac{80}{(s+1)^2} + \frac{40}{s+1}$$

Taking the inverse,
$$T_1 = 80e^{-t} + 160 \int_0^t e^{-u}\, du = 80(2 - e^{-t})$$
and
$$T_2 = 40e^{-t} + 80te^{-t} + 160 \int_0^t ue^{-u}\, du$$
$$= 40(4 - 3e^{-t} - 2te^{-t})$$
Hence after one hour $T_1 = 130 \cdot 6°$ and $T_2 = 86 \cdot 4°$.

Problem 7.6 Solve the following initial-value problems by means of Laplace transforms:

(a) $y'' + y = 1,\ y(0) = 1,\ y'(0) = 1$

(b) $y'' + y' - 2y = t,\ y(0) = 0,\ y'(0) = 1$

(c) $y'' + 2y' + 2y = \sin t,\ y(0) = 1,\ y'(0) = 0$

(d) $y'' + 2y' + 2y = f(t),\ y(0) = y'(0) = 0$

where $f(t) = \begin{cases} 2 & 0 \leqslant t < 1 \\ 0 & 1 < t \end{cases}$

(e) $y'' + y = f(t),\ y(0) = y'(0) = 0$

where $f(t) = \begin{cases} t & 0 \leqslant t < 1 \\ 0 & 1 < t \end{cases}$

(f) $y''' + 2y'' + 2y' + y = 0,\ y(0) = y''(0) = 2,\ y'(0) = 1$

(g) $\left.\begin{array}{l} x' - y' - y = 0 \\ y' + x - y = 0 \end{array}\right\} x(0) = 0,\ y(0) = 1$

(h) $\left.\begin{array}{l} x' + y' + 2x + y = 2 + e^t \\ y' + z' + y - z = 3 + e^t \\ x' + z' + 2x - z = 1 + e^t \end{array}\right\} x(0) = y(0) = z(0) = 0$

Appendix

Table A General Laplace transforms

		$\bar{x}(s)$	$x(t)$
1	Linearity	$a\bar{x}_1(s) + b\bar{x}_2(s)$	$ax_1(t) + bx_2(t)$
2	Change of scale	$\dfrac{1}{a}\bar{x}\left(\dfrac{s}{a}\right)$	$x(at)$
3		$s\bar{x}(s) - x(0)$	$\dfrac{dx}{dt} = x'(t)$
4	Differentiation	$\dfrac{\bar{x}(s)}{s}$	$\int_0^t x(u)\,du$
5	and	$\dfrac{d\bar{x}(s)}{ds}$	$-tx(t)$
6	Integration	$\int_s^\infty \bar{x}(u)\,du$	$\dfrac{x(t)}{t}$
7		$s^n \bar{x}(s) - s^{n-1} x(0) - \cdots$ $\cdots - x^{(n-1)}(0)$	$x^{(n)}(t)$
8	1st shifting theorem	$\bar{x}(s - a)$	$e^{at} x(t)$
9	2nd shifting theorem	$e^{-as}\bar{x}(s)$	$x(t - a)H(t - a)$
10	Periodic function of period T	$\dfrac{1}{1 - e^{-sT}} \int_0^T e^{-st} x(t)\,dt$	$x(t) = x(t + T)$
11	Convolution theorem	$\bar{x}(s)\bar{y}(s)$	$\int_0^t x(u)y(t - u)\,du$

Table B Particular transform pairs

	$\bar{x}(s)$	$x(t)$
1	$\dfrac{1}{s}$	1 [or $H(t)$]
2	$\dfrac{e^{-as}}{s}$	$H(t-a), a \geqslant 0$
3	$\dfrac{n!}{s^{n+1}}$	t^n $(n = 1, 2, 3, \ldots)$
4	$\dfrac{\Gamma(k+1)}{s^{k+1}}$	t^k, Re$(k) > -1$
5	$\dfrac{1}{s-a}$	e^{at}
6	$\dfrac{a}{s^2+a^2}$	$\sin at$
7	$\dfrac{s}{s^2+a^2}$	$\cos at$
8	$\dfrac{a}{s^2-a^2}$	$\sinh at$
9	$\dfrac{s}{s^2-a^2}$	$\cosh at$
10	$\dfrac{1}{(s^2+a^2)^2}$	$(\sin at - at \cos at)/2a^3$
11	$\dfrac{1}{(s^2-a^2)^2}$	$(at \cosh at - \sinh at)/2a^3$
12	$\dfrac{s}{(s^2+a^2)^2}$	$(t \sin at)/2a$
13	$\dfrac{s}{(s^2-a^2)^2}$	$(t \sinh at)/2a$
14	e^{-as}	$\delta(t-a), a \geqslant 0$

Bibliography

F. S. Acton (1970), *Numerical methods that work.* Harper International, New York.

S. Barnett and T. M. Cronin (1975), *Mathematical formulae for engineering and science students*, 2nd edition. Bradford University Press in conjunction with Crosby Lockwood Staples, London.

F. Brauer and J. A. Nohel (1969), *The qualitative theory of ordinary differential equations*, Benjamin, New York.

E. A. Coddington (1961), *An introduction to ordinary differential equations.* Prentice Hall, Englewood Cliffs, N.J.

E. A. Coddington (1955), *Theory of ordinary differential equations.* McGraw-Hill, New York.

L. Fox (Editor) (1962), *Numerical solution of ordinary and partial differential equations.* Pergamon, Oxford.

L. Fox and D. F. Mayers (1968), *Computing methods for scientists and engineers.* Oxford University Press, London.

W. Hurewicz (1958), *Lectures on ordinary differential equations*, M.I.T. Press, Cambridge, Mass.

G. N. Watson (1962), *A treatise on the theory of Bessel functions.* Cambridge University Press, Cambridge.

P. W. Williams (1972), *Numerical computation.* Nelson, London.

Answers to Problems

208   Ordinary Differential Equations

1.1

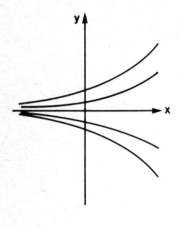

(a)

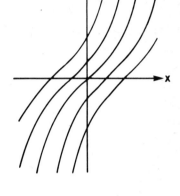

(b)

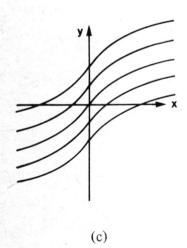

(c)

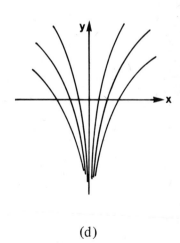

(d)

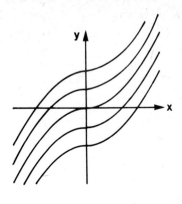

(e)

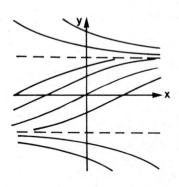

(f)

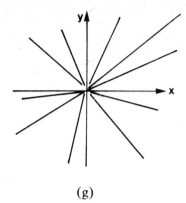

(g)

Fig. A1.1

2.1 (a) $y = ae^x$ (b) $y = a + \tan x$
    (c) $y = a + \tan^{-1} x$ (d) $y = a + \ln |x|$
    (e) $y = x^2 + a, x \geq 0, y = a - x^2, x \leq 0$
    (f) $y = (ae^x - 1)/(ae^x + 1)$ (g) $y = ax$

2.2 (a) $x = (e^t - 1)/(e^t + 1)$ (b) $v^2 = 2g[h - a \ln (a - x)/a]$
    (c) $y = 2 \sin (x^2 - \pi/2), 0 \leq x \leq \sqrt{\pi}; y = 2, x \geq \sqrt{\pi}$
    (Note that $\sqrt{(4 - y^2)} \geq 0$.)

2.3 (a) $x^2 = a \exp (y^2/x^2)$ (b) $x + 4y = axy^4$

2.4 (a) $x^2 + 2y^2 - 2xy + 2x - 6y = a$
    (b) $(y + 2x - 3)^2 (4y - x - 3) = a$
    (c) $(x - y)^2 - 2x + 10y = a$

2.5 (a) $x \tan y - e^x = a$
    (b) $x^2 \sin y + y^2 + (1 + y) \tan x = a$

2.6 (a) $y = x^2 + a/x^2$ (b) $y = (1 + x^2)^{-1/2} (a + \sinh^{-1} x)$
    (c) $y = \sin x (a - 2 \cos x)$

2.7 $y^2 = 2(1 + e^{2x} + 2x)^{-1}$

2.8 (a) $x = ae^{-p} - 4(p - 1)$, $y = a(1 + p)e^{-p} + 2(2 - p^2)$
     (b) $y = ax + a^2$ (c) $y^3 = 6a(x + a)$

2.9 $x^2 = 2ay - a^2$, $y = \pm x$

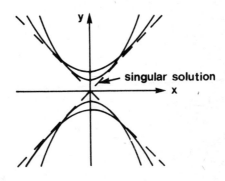

Fig. A2.9

2.10 (a) $y = a(1 + x^2)^{-1/2} \exp(k \tan^{-1} x)$
     (b) $4 \exp(-y) = a - (2x - 1) \exp(2x)$
     (c) $y = x \ln(a + \ln x)$
     (d) $a = 8y - 4x + 5 \ln(1 + 4x + 8y)$
     (e) $a = x \cos y + x^2 e^x$
     (f) $y = x \csc x \, (a - 1/x - 1/2x^2)$
     (g) $5xy = (a - \cos 5x) \exp(-x^2)$
     (h) $1/y = ae^x - 2 \sin x$
     (i) $y = ax - a^2$
     (j) $y = a(a - x)^2$

2.11 (a) $(3y - x)(y + x) = 4$ (b) $y = x^2 - 1 + 2x \ln x$
(c) $y^{-4} = 1 - 4x - \exp(-4x)$

3.1 (a) 0·603 55, 0·638 07 (b) 0·628 42, 0·636 71
(c) 0·631 38, 0·636 65   exact 0·636 62

3.2 0·636 71   **3.3** $(16A_1 - A_0)/15$, 0·636 62

3.4 $\alpha + \beta + \gamma = 1$, $\beta a + \gamma c = \beta b + \gamma d + \gamma e = \frac{1}{2}$,
$\beta a^2 + \gamma c^2 = \beta b^2 + \gamma d^2 + 2\gamma de + \gamma e^2 = \beta ab + \gamma cd + \gamma ce = \frac{1}{3}$
$\gamma ea = \gamma eb = \frac{1}{6}$

3.5 1·242 67, 1·242 79, 2×10^{-5}

3.6 (a) 1·5839, -2×10^{-4} (b) 1·583 65, -3×10^{-7}

3.7 $\frac{251}{720}h^5 f_0^{(4)}$, $-\frac{19}{720}h^5 f_0^{(4)}$, $\bar{y}_1 = \bar{y}_1^c - \frac{19}{270}(\bar{y}_1^c - \bar{y}_1^p)$

3.8 1·583 65, -7×10^{-7}

4.1 (a) Fig. A4.1(a), oscillatory (b) Fig. A4.1(b), divergent

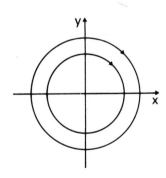

Fig. A4.1a

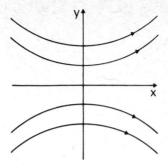

Fig. A4.1b

4.2 (a) Fig. A4.2(a), decayed oscillation
(b) Fig. A4.2(b), divergent

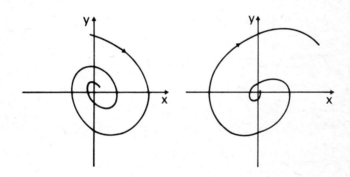

Fig. A4.2a–b

4.3 (a) (i) $x = Ae^{-t} + Be^{4t}$ (ii) $x = e^{-t/2}(A \sin \tfrac{1}{2}t + B \cos \tfrac{1}{2}t)$
(iii) $x = (A + Bt)e^{3t/2}$
(b) (i) $x = e^{-t}$ (ii) $x = 2e^{-t/2} \sin \tfrac{1}{2}t$
(iii) $x = [1 - (1 - e^{-3/2})t]e^{3t/2}$

4.4 (a) $-\frac{1}{2}t + \frac{1}{4}$ (b) $-\frac{1}{2}e^t$ (c) $-\frac{1}{3}te^{-t}$
(d) $-\frac{3}{10}\sin t + \frac{1}{10}\cos t$ (e) $-e^{-t}(\frac{1}{6}t^2 + \frac{1}{9}t)$

4.5 $\frac{1}{8}e^t(\sin t - \cos t)$

4.6 (a) $x = A + Bt + Ce^{2t} - t^2$
(b) $x = Ae^t + Be^{-t} + C\sin t + D\cos t + \frac{1}{15}\sin 2t$
(c) $x = Ae^{-t} + e^{-t}(B\sin t + C\cos t)$

4.7 (a) $x = -3t - 2, y = 3t^2 + 8t + A$
(b) $x = A\cos t + B\sin t; y = \frac{1}{2}(A+B)\cos t + \frac{1}{2}(B-A)\sin t$
(c) $x = Ae^{-7t/5} + \frac{3}{14}t - \frac{1}{98}, y = 3Ae^{-7t/5} + \frac{1}{7}t - \frac{26}{49}$
(d) $x = Ae^{-2t} + \frac{1}{6}e^t, y = Be^{-t} + 2 + \frac{1}{4}e^t, z = Ce^t - 1 + \frac{1}{2}te^t$

4.8 $-1, \begin{bmatrix} A \\ -A \end{bmatrix}$ $2, \begin{bmatrix} B \\ -4B \end{bmatrix}$

$x = Ae^{-t} + Be^{2t}, y = -Ae^{-t} - 4Be^{2t}$

4.9 $x = V\cos\alpha t, y = V\sin\alpha t - \frac{1}{2}gt^2; y = \tan\alpha x - (g\sec^2\alpha/2V^2)x^2$

4.10 (a) $x = t - \ln(1 - Ae^{2t}) + B$ (b) $x = At^3 + 18A^2t^2 + Bt + C$
(c) $2Ax = \tan(At + B)$, or $x = C$
(d) $x = At + t(B\cos\ln t + C\sin\ln t) + t\ln t$

4.11 (a) $x = e^{-2t}(A\cos 3t + B\sin 3t) + \frac{1}{169}(13t - 4) + \frac{1}{10}(3\sin t - \cos t)$
(b) $x = (A + Bt + Ct^2)e^{2t} + \frac{1}{24}t^4 e^{2t}$
(c) $x = \frac{1}{3}t - \frac{13}{6}Ae^{-3t/2} + B, y = \frac{3}{2}Ae^{-3t/2}, z = \frac{1}{3} + Ae^{-3t/2}$
(d) $x = At + Bt\ln t + \frac{3}{2}t(\ln t)^2 + 1$

5.2 (a) $0.7967, -1.1848$
(b) $0.7971, -1.1862; -6 \times 10^{-5}; 2 \times 10^{-4}$

5.3 (a) and (b) $0.0913, 0.4193$

5.4 $\bar{z}_1^p = 0.371\,255, \bar{z}_1^c = 0.371\,334, \bar{y}_1^c = 0.315\,776$

5.5 2.7698 **5.6** 0.2633

6.1 (a) $x - \frac{x^3}{3!} + \frac{x^5}{5!} - \cdots + \frac{(-1)^n x^{2n+1}}{(2n+1)!} + \cdots$

(b) $1 - x + \dfrac{x^3}{3} - \dfrac{x^4}{6} + \dfrac{x^5}{30} - \cdots + \dfrac{2^{n/2} \cos n\pi/4 \cos n\pi x^n}{n!} + \cdots$

6.2 (a) $1 + x - \dfrac{x^2}{2} - \dfrac{x^4}{4!} - \dfrac{3x^6}{6!} - \cdots - \dfrac{(2n-3)!x^{2n}}{2^{n-2}(n-2)!(2n)!} - \cdots$ all x

(b) $1 - \dfrac{3}{2^2 \, 2!} x^2 - \dfrac{3 \cdot 1 \cdot 3 \cdot 7}{2^4 \, 4!} x^4 - \cdots$,

$$A_n = \dfrac{(2n-5)(2n-1)}{4n(n-1)} A_{n-2}; \; |x| < 1$$

6.3 (a) 0·0998, 0·9950   **(b)** 0·9003, −0·9907
    (c) 1·0950, 0·8998   **(d)** 0·9962, −0·0757

6.4 (a) $y = Ax + B\left(1 - \dfrac{x^2}{2} - \cdots - \dfrac{(2n-3)!x^{2n}}{2^{n-2}(n-2)!(2n)!} - \cdots\right)$

(b) $y = A\left(1 - \dfrac{3}{2^2 \, 2!}x^2 - \dfrac{3 \cdot 1 \cdot 3 \cdot 7}{2^4 \, 4!}x^4 - \dfrac{7 \cdot 3 \cdot 1 \cdot 3 \cdot 7 \cdot 11}{2^6 \, 6!}x^6 - \cdots\right)$

$\qquad + B\left(x + \dfrac{5}{2^2 \, 3!}x^3 + \dfrac{5 \cdot 1 \cdot 5 \cdot 9}{2^4 \, 5!}x^5\right.$

$\qquad \left. + \dfrac{9 \cdot 5 \cdot 1 \cdot 5 \cdot 9 \cdot 13}{2^6 \, 7!}x^7 + \cdots\right)$

6.5 $y = \dfrac{A}{x} + B\left[1 + \dfrac{1}{2x^2} - \cdots + \dfrac{(-1)^n 1 \cdot 3 \cdot 5 \cdots (2n-3)}{2^n n! x^{2n}} + \cdots\right]$
$\quad |x| \geq 1$

6.6 (a) $0 < |x| < 1$ **(b)** $0 < |x| < \tfrac{1}{2}$ **(c)** $0 < |x| < 1/\sqrt{2}$

6.7 (a) $y = A(1 + 5x + \tfrac{7}{2}x^2 + \tfrac{65}{42}x^3 + \cdots) + Bx^{2/3}\left(1 + x + \dfrac{x^2}{2!} + \dfrac{x^3}{3!} + \cdots\right)$
     all x

(b) $y = A/x + B \displaystyle\sum_{n=0}^{\infty} x^n, \; 0 < |x| < 1$

(c) $x = (A + B \ln t)\left(-\dfrac{t^4}{2^3 2!} + \dfrac{t^6}{2^5 3!1!} - \dfrac{t^8}{2^7 4!2!} + \cdots\right)$

$\quad + B\left[1 + \dfrac{t^2}{2^2} + \dfrac{t^4}{2^5 2!} - \dfrac{(1 + \frac{1}{2} + \frac{1}{3})t^6}{2^6 3!1!}\right.$

$\quad + \dfrac{[(1 + \frac{1}{2} + \frac{1}{3} + \frac{1}{4}) + \frac{1}{2}]t^8}{2^8 4!2!} - \cdots\Big], \; 0 < |t|$

(d) $x = (A + B \ln t)\left(t - \dfrac{t^3}{2^2} + \dfrac{t^5}{2^4 (2!)^2} - \cdots\right)$

$\quad + B\left[\dfrac{t^3}{2^2} - \dfrac{(1 + \frac{1}{2})t^5}{2^4 (2!)^2} + \cdots\right], \; 0 < |t|$

6.15 (a) $y = 1 + x - \frac{1}{2}x^2 + \frac{1}{2}x^3$

(b) $y = A[1 - x + \frac{1}{2}x^2 + \frac{1}{6}A^{-2}(2 - A^2)x^3]$

(c) $y = A(1 - \frac{1}{6}x^3 + \frac{1}{180}x^6 - \cdots) + B(x - \frac{1}{12}x^4 + \cdots)$, all x

(d) $y = (1/x^2)(A \cos x + B \sin x)$

(e) $x = e^t(A + B \ln t)$ (f) $x = At + B(\frac{1}{2}t^{-1} + t \ln t)$

6.16 $y = [A\sqrt{(x - 1)} + B(x - 1)][1 - (x - 1) + (x - 1)^2 - \cdots]$

6.17 $y = x^3 [A \sinh (1/x) + B \cosh (1/x)]$

7.2 (a) $2/(s + 3)$ (b) $6/s^4 - 1/(s + 1)^2$ (c) $5s/(s^2 + 9)$
(d) $720/s^7 + 21/s^4 + 1/s$ (e) $2/(s^2 - 2s + 5)$

7.3 (a) Fig A7.3(a) $k/s^2 T - ke^{-sT}/s(1 - e^{-sT})$
(b) Fig A7.3(b) $\omega(1 + e^{-s\pi/\omega})/(s^2 + \omega^2)(1 - e^{-s/\pi\omega})$

7.4 (a) $t - (t - 1)H(t - 1), \; (1 - e^{-s})/s^2$
(b) $\sin t[1 - 2H(t - \pi) + H(t - 2\pi)], \; (1 + 2e^{-\pi s} + e^{-2\pi s})/(s^2 + 1)$
(c) $t - 2(t - a)H(t - a) + 2(t - 2a)H(t - 2a) - \cdots$,
$1/s^2(1 - 2e^{-as} + 2e^{-2as} - \cdots) = (1 - e^{-as})/s^2(1 + e^{-as})$
(d) $te^{-t} + (1 - te^{-t})H(t - 2)$,
$1/(s + 1)^2 + e^{-2s}/s - e^{-2(s+1)}/(s + 1)^2 - 2e^{-2(s+1)}/(s + 1)$

7.5 (a) $\frac{1}{2} \sin 2t$ (b) $3e^{-t}$
(c) $\cos \sqrt{3}t$ (d) $t + \frac{1}{2}t^2 - \cos 3t + \frac{1}{3} \sin 3t$
(e) $e^{-2t}(\cos \sqrt{5}t - (3/\sqrt{5}) \sin \sqrt{5}t)$

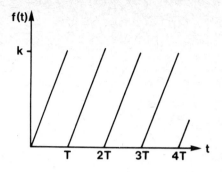

Fig. A7.3a

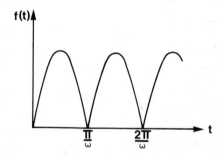

Fig. A7.3b

(f) $e^{t/2} [\cos(\sqrt{11}t/2) + (1/\sqrt{11})\sin(\sqrt{11}t/2)]$
(g) $t - 1 + e^{-t}$ \qquad (h) $e^{-t}(2-t) - 2e^{-2t}$
(i) $\frac{1}{3}[e^t + (\sqrt{2}/2)e^{-t}\sin\sqrt{2}t - e^{-t}\cos\sqrt{2}t]$
(j) $e^{-2t}(t - \sin t)$ \qquad (k) $\frac{1}{8}t(\sin t - t\cos t)$
(l) $2e^t\sqrt{(t/\pi)}$ \qquad (m) $H(t-1)$
(n) $(t-2)e^{-(t-2)}H(t-2)$ \qquad (o) $-\cos t\, e^{-(t-\pi)}H(t-\pi)$
(p) $[t - 2 + e^{-(t-1)}]H(t-1)$

7.6 (a) $1 + \sin t$ (b) $\frac{1}{12}(8e^t - 5e^{-2t} - 6t - 3)$
(c) $\frac{1}{5}[(7\cos t + 6\sin t)e^{-t} + \sin t - 2\cos t]$
(d) $1 - (\cos t + \sin t)e^{-t} - \{1 - [\cos(t-1) + \sin(t-1)]e^{-(t-1)}\} H(t-1)$
(e) $t - \sin t - [t \sin(t-1) - \cos(t-1)] H(t-1)$
(f) $5e^{-t} + 3[\sqrt{3} \sin(\sqrt{3}t/2) - \cos(\sqrt{3}t/2)] e^{-t/2}$
(g) $x = 2\sin t, y = \sin t + \cos t$
(h) $x = \frac{1}{6}(e^t - e^{-2t}), y = \frac{1}{4}(8 + e^t - 9e^{-t}), z = \frac{1}{2}[(2+t)e^t - 2]$

Index

Adams-Bashforth-Moulton method 77
Algebraic method 124
Autonomous equation 79

Bernoulli's equation 46
Bessel function 159
Bessel's equation 158
Bibliography 206
Boundary conditions 2
Boundary value problems 121

Characteristic equation 85
Clairaut's equation 49
Complementary function 88
Constants of integration 28

Degree 17
Delta function 182, 192
Dependent variable 16
Determinant
   second order 101
   third order 103
Differential equations
   analytic solution 2
   approximate solution 2
   autonomous 79
   Bernoulli 46

Bessel 158
Clairaut 49
classification 16
degree 17
Euler's linear 110
exact 40
first order linear 43
formulation 4
general solution 18
homogeneous first order 36
homogeneous nth order 85
homogeneous second order 84
Legendre 158
linearity 16
numerical solution 3
order 16
ordinary 1
partial 1
particular solution 18
separable 33
singular solution 22, 23, 49
solution 2
stiff 126

Eigenvalue 106
Eigenvector 106
Envelope 22, 50

Index

Euler's linear equation 110
Euler's method 28, 68
Exact equation 40
Existence of solution 30

First order linear equation 43

Global error 60

Heaviside unit step function 173, 177
Homogeneous equation
  first order 36
  nth order 85
  second order 84

Ill conditioning 73
Independent variable 16
Indicial equation 144
Initial conditions 2
Initial value problems 114
Integrating factor 43
Inverse Laplace transform 184

Laplace transforms
  convolution theorem 177
  definition 166
  inverse 184
  rules for inversion 184
  rules for transforms 168
  tables 204, 205
Legendre polynomial 161
Legendre's equation 158
Leibnitz theorem 132
Linear equations 43, 84, 88, 99, 100, 190
Local error 60

Maclaurin series 130, 134
Mathematical model 6
Mid point rule 57
Milne's method 70

Numerical methods of solution
  Adams-Bashforth-Moulton 77
  algebraic 124
  Euler 28, 68
  mid point rule 57
  Milne 70
  one-step 62
  Picard 29, 31
  predictor-corrector 68, 118
  Runge-Kutta 63, 115
  series 63, 114, 121
  Simpson's rule 57
  trapezoidal rule 57
  trial and error 122

One-step methods 62
Ordinary point 131

Parasitic solution 75
Particular integral 88
Phase space 24, 79, 80
Picard's method 29, 31
Predictor-corrector methods 68, 118

Regular singular point 140
Riccati transformation 127
Richardson's extrapolation 61
Runge-Kutta 63, 115

Separable equation 33
Series solution 63, 114, 121, 130
Simple harmonic motion 27
Simpson's rule 57
Singular point 131
Singular solution 22, 23, 49
Solution curve 17
Stability 75
Steady state solution 45
Stiff equation 126

Taylor series 59, 63, 130, 134
Trapezoidal rule 57
Trial and error method 122

Undetermined coefficients 89
Uniqueness of solution 30

$$a_0 = \frac{2}{T_1}\int_0^{2\pi} \cos\frac{x}{2} = \frac{1}{T_1}\left[2\sin\frac{x}{2}\right]_0^{\pi \cdot 2\pi}$$